Chevy LS1/LS6 Performance

High Performance Modifications for Street and Racing. Covers Chevy LS1 and LS6 Engines, 1997 and Up

Christopher P. Endres

HPBOOKS

HPBooks
Published by the Penguin Group
Penguin Group (USA) Inc.
375 Hudson Street, New York, New York 10014, USA
Penguin Group (Canada), 90 Eglinton Avenue East, Suite 700, Toronto, Ontario M4P 2Y3, Canada
(a division of Pearson Penguin Canada Inc.)
Penguin Books Ltd., 80 Strand, London WC2R 0RL, England
Penguin Group Ireland, 25 St. Stephen's Green, Dublin 2, Ireland (a division of Penguin Books Ltd.)
Penguin Group (Australia), 250 Camberwell Road, Camberwell, Victoria 3124, Australia
(a division of Pearson Australia Group Pty. Ltd.)
Penguin Books India Pvt. Ltd., 11 Community Centre, Panchsheel Park, New Delhi—110 017, India
Penguin Group (NZ), 67 Apollo Drive, Rosedale, North Shore 0632, New Zealand
(a division of Pearson New Zealand Ltd.)
Penguin Books (South Africa) (Pty.) Ltd., 24 Sturdee Avenue, Rosebank, Johannesburg 2196, South Africa

Penguin Books Ltd., Registered Offices: 80 Strand, London WC2R 0RL, England

CHEVY LSI/LS6 PERFORMANCE

Cover design by Bird Studios
Cover illustration by David Kimble, courtesy of Chevrolet
Book design and production by Michael Lutfy
Interior photos by author unless otherwise noted

First edition: April 2003

ISBN: 978-1-55788-407-7

PRINTED IN THE UNITED STATES OF AMERICA

20 19 18 17 16 15

CONTENTS

ACKNOWLEDGMENTS

A wise editor once told me, "Chris, you're a writer. You're not supposed to know everything, you're supposed to know everyone." That said, there is a long list of people that contributed significantly to the creation of this book. To them, I owe a huge debt of gratitude. They are:

Nick Agostino, Wade Stevens, Dave DiLuca, and Barry Dwyer, Agostino Racing Engines;
Jayson Cohen, Motorsport Technologies Inc.;
Cameron Evans, *Popular Hot Rodding* magazine;
Johnny Hunkins, *GM High-Tech Performance* magazine;
Tom Izzo, Speed Inc.;
James Jureski, Texas Nitrous Technology;
John Juriga and Tom Read, GM Powertrain;
Jonathan Larson;
Steve Lock;
Rob Raymer, LS1 Motorsports;
Doug Rippie, Doug Rippie Motorsports;
Rob Simoes, CorvetteCavalry.com;
John Skiba, Pace Performance Parts;
Kevin Woodruff, SLP Performance Parts;

And most importantly, my chief proofreader, enthusiastic motivator, and dear wife: Mary. LTS

INTRODUCTION

I can tell you the precise instant it happened; the moment I became obsessed with the LS1. My sister pulled up in a beautiful Sebring Silver Z28 that she was test driving. While I wasn't too sure about the newly restyled front end, I agreed to test drive it for her. As a devout EFI car guy, I never believed the new LS1 engine could measure up to the LT1-powered Z28, which had been serving as my current project. I will never forget (nor will my sister, I'm sure) the test drive we took in that six-speed Camaro.

From that early spring day in 1998, I became increasingly fascinated by the LS1. I just couldn't wrap my mind around the power made by a brand-new car that was completely bone stock—not a single modification. The following summer, that very same Z28 changed hands and took up residence in my garage. Never one to leave well enough alone, I delved right in to the modification of my new-age performer. In the years that followed, I studied every nuance of the engine in pursuit of maximizing its potential. This book is the culmination of all I've learned about the LS1. I hope you find it helpful in your own pursuit of power. —*Chris Endres*

1 Meet the LS1

Of RPO Codes and Other GM-speak

The saying goes "What's old is new again" and nowhere is this more true than at General Motors Corporation. GM has a long and rich tradition of transforming its three-character Regular Production Option (RPO) codes into model names. Though alpha-numeric codes such as WS6, Z28, ZR1, and Z06 came into existence as obscure RPO designators, they have since become legendary in automotive enthusiasts' circles.

Engine designators often take on a life of their own too: L-88, LT-1 (not to be confused with the fuel-injected and non-hyphenated LT1 of the mid '90s!), and the LT5 are just a few of the better-known examples. Through the late-'90s and into the early '00s, the LS1 and LS6 engines have become two more RPO's matching or even overshadowing the cars they power. RPO code LS1 is GM-speak for the Generation-III, medium displacement 5.7L sequential port fuel-injected V8 engine. For more than a year before the all-new 1997 Corvette was unveiled, it had been widely reported that C5 (more GM-speak, for fifth-generation Corvette) would be the first General Motors product to use an engine from this new family of medium-displacement V8's. The story by which these engines came into being is an interesting one.

In 1991, the fuel-injected LT1 (no hyphen!) was just about to make its debut in the 1992 Corvette, replacing the venerable Tuned Port Injected 5.7-liter L98. While General Motors touted LT1's new cylinder heads, intake manifold, reverse-flow cooling and optical distributor as revolutionary, the real revolution was just gaining momentum deep within GM's Powertrain Division. Literally tens of millions of small-block V8's had been produced since the engine's appearance in 1955. And though its fundamental design was excellent, its limitations had been reached on the way to addressing the increasingly stringent emissions regulations, tougher fuel efficiency standards, and

The venerable LT1 engine powered the Corvette from 1992–96 and the F-body from 1993–97. This Gen-II small block proved a very capable performer.

most importantly: increasing customer demands for a truly world-class performance engine.

Generation-III Is Born

Feasibility studies began at Powertrain during the winter of 1991 to determine the viability of a Generation-III small-block V8 engine. The initial plan was to improve the existing engine rather than create an entirely new one. According to GM Powertrain Assistant Chief Engineer for the Gen-III, John Juriga, "The Flint engine assembly plant (where the Gen-II engine was built) and its tooling had outlived its usefulness. An efficiency study showed us that it would be cheaper to open a new plant elsewhere than retool Flint. This opened the door to allow us to do something completely new. Our goals for performance and durability were clearly beyond the Gen-II's capabilities." Those goals included:

- Increased power and torque
- Improved fuel efficiency
- Reduced emissions
- Compact and lightweight
- Refined Noise, Vibration and Harshness characteristics
- Enhanced quality, reliability and durability

By February of 1993, Powertrain declared its intention to discontinue the Gen-III project as a small block, and instead would design and create a completely new medium-displacement V8 engine.

Juriga said, "We had very specific goals for the modularity of this engine. Because Gen-III was to be a complete replacement for the old small block everywhere it was used, it was essential that it be a workhorse, not a specialty engine. It was our job to make sure we could build all variants of the new engine on the same assembly line. The budget for Gen-III was over $1.2 billion, which is a very large program, even for GM."

The power goal for Gen-III was to produce 1 horsepower and 1 lb.-ft. of torque per cubic inch of displacement (for Corvette-bound LS1s). Weight reduction was another important consideration; a sixty-pound decrease over Gen-II was deemed necessary. Gen-III's major fuel economy initiatives were focused on friction reduction. Objectives regarding durability, reliability, and driveability were simply (and ambitiously): Best in class.

The minimization of Noise, Vibration and Harshness (NVH) was an important priority for Gen-III, as an enhanced element of refinement was required for the new Corvette. As GM's flagship automobile, Corvette traditionally receives GM's newest and best technology. Development of the Gen-III was fortunate to occur at the same time Chevrolet was completely rethinking the Corvette. The concurrence of these two events presented a unique opportunity to pair a brand-new platform with a completely new drivetrain, and allowed the car and its powerplant to be integrated perfectly.

The early prototypes of the Gen-III first ran on the GM Powertrain dynamometers sometime late in the winter of 1993 and passed the Concept Initiation "gate" of the development process in April of 1993. Passing this gate meant that the new engine had demonstrated the capability to meet the requirements set for the project. Concept Approval was granted in May 1994, when the LS1 was deemed within 10% of all development targets.

The C5 alpha build began in June 1994, producing the first of the

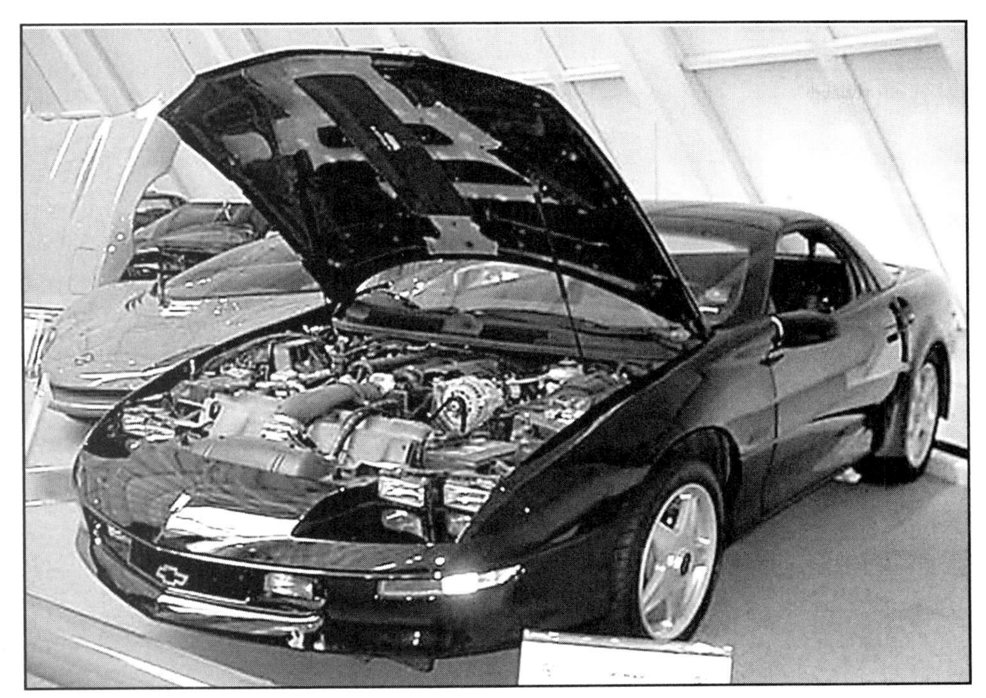

In 1995, "Cormaros" were employed as stealth test vehicles. (Photo courtesy of CorvetteCavalry.com)

C5 Beta cars roamed the streets of Detroit and deserts of Arizona starting in 1995. They wore odd cloaking to keep the spy photographers guessing at the true shape of the car beneath. (Photos courtesy of CorvetteCavalry.com)

prototype 1997 Corvettes, but there were no Gen-III engines to power them. Porosity problems hampered the early aluminum blocks, resulting with oil in the coolant and vice-versa, and caused numerous delays in testing. By late summer, the C5 alpha phase was in high gear and most of the cars had iron block/aluminum head Gen-III's installed. As the year progressed, all-aluminum units showed up and proved more reliable.

The Cormaro

During the first half of 1995, Powertrain came up with an interesting way to put street miles on the Gen-III powerplant without drawing unwanted attention. GM equipped twenty 1995 Camaro body shells with C5 chassis and powertrain combinations and dubbed them the "Cormaro." It made sense for GM to use the F-body platform as it had become traditional that they receive a slightly detuned "Corvette Engine" the year after it debuts in the 'Vette. A large number of C4's were also fitted with Gen-III engines to develop cold-start calibration and traction control system software.

The beta-build C5s were quite representative of production cars, and appeared toward the end of June 1995. By autumn of 1995 the beta build was completed, but the number of cars in testing sometimes exceeded the supply of reliable, all-aluminum engines. Where vehicle mass approximation was not critical, betas were sometimes fitted with iron block/aluminum head combinations. By late 1995 or early 1996, the aluminum block shortage was less of a problem and the Gen-III progressed to final development in prototype C5s. This phase of final development continued into April of 1996 when the engine was released for production.

In the spring of 1996, anticipation in the Corvette community reached a fevered pitch. While the '96 Corvette was an exceptional performer with its optional LT4, rumors were running rampant about the new aluminum engine for 1997. While skeptics felt it unlikely that the new engine would perform on par with the 330 hp LT4, GM wasn't quite ready to give away the surprise.

In June 1996, nearly eight months before the introduction of the 1997 Corvette, GM had decided that the time had come to unveil the new V8, now known as the LS1, at a technical

The LT4 engine was a hopped-up version of the LT1 available in all six-speed Corvettes in 1996, and a handful of F-bodies in '97. It was rated at 335 horsepower.

The LS1 represents a major breakthrough in so-called old tech pushrod engines. Compact dimensions, light weight and prodigious power combine to make it one of the best V8's ever conceived.

media presentation in Los Angeles. This was not the first time GM had elected to introduce a new Corvette engine before its car. In 1988, GM presented the automotive media with a comprehensive seminar on the LT5 nine months before showing the ZR1 it was to power.

The LS1

The new engine was a tremendous step forward not only in architecture, but also in material utilization and process control. As the foundation of the engine, the cylinder case set a precedent for increased stiffness and reduced vibration that was carried throughout. Notable elements of its design include deep bore skirts that extend past the crank centerline and the main bearing caps. The main caps feature six-bolts, which includes two horizontal bolts fortifying the conventional four vertical fasteners. Lightweight pistons, powder metal connecting rods and a revised firing order allow for smooth high rpm power production.

The LS1 oil pan is an integral part of the lower engine structure and contributes to overall cylinder case rigidity. The Corvette has a unique pan that incorporates extended sumps to ensure adequate oil supply to the pickup under all operating conditions. Sources involved in C5 development call the unusual-looking aluminum casting the "gull wing" pan. F-body and truck oil pans are structural, too, but forego the extended sump. A small, PF-44 oil filter mounts on the rear of the oil pan rather than the block. Though no oil cooler is available, extensive testing had shown the oil temperature comparable to that of the Gen-II small block and therefore the cooler was deemed unnecessary. Where feasible, sealing surfaces are in a single plane and utilize molded-in-place or carrier-style gaskets. Sealing surfaces and sensor locations are situated away from submerged areas. In lieu of pressed-in water jacket plugs, the engine block uses threaded plugs to minimize the chance of fluid leaks.

The LS1's lightweight aluminum cylinder heads feature "replicated" ports and a super-efficient combustion chamber design that optimizes airflow into the engine. This is one rare place that modularity of the LS1 has a disadvantage. Juriga said: "The Gen-III valve center dimensions are common for all bore sizes. When you

Corvette was the first GM vehicle to use Electronic Throttle Control or "Drive by Wire." This innovation de-coupled the mechanical control of the throttle pedal over the throttle body, giving it instead to the car's PCM. Functions such as traction control and active handling are much more seamless and precise as a result.

design for single bore, you can optimize the valve center separately for each bore, optimizing power."

The valvetrain features planar valve, rocker arm and pushrod geometry for reduced stress and friction. The cylinder head sealing system has been improved with a new block, head and head gasket design.

The plenum of the LS1's composite intake manifold occupies virtually all of the space available in the valley between the cylinder heads, minimizing the overall height of the engine. This arrangement allows the runners a long, smooth curve from the plenum, up and over to each intake port in the heads. No oil or water comes in contact with the intake manifold, further reducing the likelihood of an annoying leak. Fuel injector targeting is optimized, aiming fuel directly at the backside of the intake valves.

Gone is the reverse-flow coolant path of the Gen-II. The cooling passages around the combustion chambers are more efficient than those of its predecessor, and when combined with improved anti-friction technology in the block, pistons and rings; it made sense to revert to a traditional coolant flow path. This also served to add robustness to the cooling system, particularly in the area of de-aeration.

Sequential Fuel Injection aids combustion efficiency by optimizing fuel delivery through a mass airflow sensor and one ignition coil per cylinder while OBD-II helps reduce emissions. For the Corvette, Electronic Throttle Control (ETC) allows precise tailoring of a specific throttle input. ETC integrates the functionality of electronic speed control, brake torque management and traction control into a single controller, for mass reduction and high durability. F-body and truck applications utilize conventional cable-actuated throttles.

Designing very rigid accessory mounts reduces noise and vibration inherent in these systems. New, quieter running accessories and a drive system that employs two serpentine belts further reduce noise. One belt drives the air-conditioning compressor while the other drives the remaining accessories. The dual serpentine belts decouple the generator and air-conditioning compressor for improved noise isolation and represent a 7 dB(A) reduction in accessory drive noise at 1000 rpm. The LS1 is equipped with an AC-Delco CS-series alternator, a Gerotor power steering pump, an all-new water pump, and a new air-

The LS1 is a remarkably quiet engine; thanks in part to intelligent accessory drive design. The air-conditioning compressor has been decoupled from the other accessories, substantially reducing noise.

Application*	1997	1998	1999	2000	2001	2002	2003
			Model Year				
LS1 5.7 Corvette	345/350	345/350	345/350	345/350	350/375	350/375	350/375
LS6 5.7 Corvette Z06	—	—	—	—	385/385	405/400	405/400
LS1 5.7 Base F-body	—	305/335	305/335	305/335	310/340	310/340	—
LS1 5.7 Ram Air F-body	—	320/345	320/345	320/345	325/350	325/350	—
LR4 4.8 Truck	—	—	255/285	270/285	270/285	275/290	275/290
LM4 5.3 Truck	—	—	—	—	—	—	290/330
LM7 5.3 Truck	—	—	270/315	285/325	285/325	285/325	295/330
L59 5.3 Flex Fuel Truck	—	—	—	—	—	285/325	285/325
LQ4 6.0 Truck	—	—	300/355	300/355	300/360	325/370	325/370
LQ9 6.0 Truck	—	—	—	—	—	345/380	345/380

* For engines in varying applications, highest output version is listed.

The LS1 debuted in the F-body in 1998, making LS1-equipped Camaros and Firebirds the best bang for the buck musclecar available.

conditioning compressor. Direct mounting of these accessories allows for fewer bolts, fasteners and attachment points. The utilization of stiffer components and longer fasteners that thread deep into the backbone of the engine ensures adequate clamp load over time.

The Numbers

The LS1 easily achieved Powertrain's goals for both power and torque. For model year 1997, the LS1 generated 345 horsepower at 5600 rpm and 350 lb.-ft. of torque at 4400 rpm, with a maximum engine speed of 6200 rpm. When compared to the 1996 LT1, these figures represent increases in peak power of 15%, peak torque of 5%, while achieving mass reduction of 12%, and brake specific fuel consumption improvements of 4%. The LS1 generated an additional 15 horsepower at a 200 rpm lower peak, than the 1996 LT4. It also produces 10 more lb.-ft. of torque, with that peak coming 100 rpm lower than the LT4. If you were to rate the '97 LS1 using the gross power method of the 1960s it would put out about 390–400 horsepower.

For model year 1998, the LS1 went largely unchanged in the Corvette, so the big news was its availability in the F-bodies. Though the engine was quite similar, it was rated a very conservative 305 horsepower when installed in the Camaro Z28, Firebird Formula or Trans Am. Power was bumped to 320 if the car was optioned as an SS or WS6 Ram Air model. Notable changes made for fitment in the F-bodies were a conventional cable actuated throttle body, a

Corvette convertibles receive the same power train as the standard coupes.

The Z06 is the ultimate Corvette. It embodies all of the traits that make the LS1 so great: lightweight, responsive and refined.

traditionally shaped (but still structural) oil pan, and more restrictive intake and exhaust systems, which were mandated by packaging constraints.

1999 Model Changes—Things were mostly status quo for the LS1 cars in 1999, though both platforms received smaller fuel injectors. This did not impact rated power output, but can slightly limit performance when modifications are performed. The Gen-III engine debuted in the all-new Silverado and Sierra light trucks with 4.8-, 5.3- and 6.0-liter variants. The two smaller engines were configured as iron block and aluminum head combinations, while the 6.0-liter LQ4 had iron block and heads. Of course, these engines were modular as well, in that most components were interchangeable between the truck and car applications.

2000 Model Year—Model year 2000 changes included cast iron exhaust manifolds for the F-body, and a slightly more aggressive camshaft for Corvette. The larger cam compensated for more restrictive exhaust with the addition of two small "pup" catalytic converters. This allowed the C5 to maintain its 345 horsepower rating while improving emissions.

The Corvette Z06 and LS6 Engine

Things got really interesting in 2001 with arrival of the Corvette Z06, and with it the return of RPO code LS6. (Equally as traditional at GM is the recycling of model names and RPO codes.) When news of the forthcoming LS6 first broke, many optimistic enthusiasts mistakenly believed Chevrolet was to resume installation of the hallowed 454 cubic-inch big-block in the Corvette. While the notion was intriguing, the fact is that the Gen-III makes as much power, while weighing less, and using far less fuel than any big block ever could. GM rated the LS6 in the 2001 Corvette Z06 at 385 horsepower and 385 lb.-ft. of torque. This is 35 horsepower and 25 lb.-ft. of torque more than an LS1-powered Corvette of the same vintage.

The best feature of the Z06 was the LS6 engine. More evolutionary than revolutionary, the LS6 was an updated version of the LS1 with even more emphasis placed on performance. The LS6 block incorporated bulkhead windows, which allowed better crankcase breathing between bays and resulted in decreased crankcase air pressure. The updated composite intake manifold featured increased plenum volume and improved intake runners to deliver more air to the combustion chamber with less turbulence. The LS6 cylinder heads feature improved ports and combustion chambers. A revised camshaft takes advantage of the LS6 engine's newfound inhalation ability. Fuel injectors were upgraded, flowing 10% more fuel than those found on the LS1. A larger free-flowing mass airflow sensor with integrated intake

Z-OH!-6

When Chevrolet reintroduced the long-absent fixed roof coupe (FRC) to the Corvette lineup for the 1999 model year, it was touted as the "purist's Corvette." By limiting the available options, mandating a six-speed gearbox and Z51 sport (read: autocross) suspension, the fixed roof coupe was definitely not intended for the weekend recreational cruiser. It was a car designed for competition: road racing and autocross, specifically. So from the hardcore enthusiast's point of view, the fixed roof coupes were a roaring success. But with just over 6100 examples turned out in their two-year run, Chevrolet likely didn't see the model in the same favorable light. The lack of options, notably limited paint choices, was most often blamed for sluggish sales. After an inaugural run of 4031 cars produced for model year 1999, sales crumbled in the car's sophomore year, with just 2080 FRC's hitting the streets.

GM had clearly put itself in a difficult position. Presumably, they had made a substantial investment in design and tooling to make the hardtop coupe a reality. Financially, it didn't make any sense to abandon the body style after just two model years. What to do? Enter the Corvette Z06. The added stiffness and reduced mass of the fixed roof coupe made it a natural platform for a new performance package, and henceforth, all 2001 and newer Corvette hardtops are Z06's, and vice versa.

air temperature sensor was added, as well. Red fuel rail covers are the only outward clue to alert the casual observer that the engine before them is an LS6.

GM made the most of the LS6's additional high-rev capability by raising the engine's redline to 6500 rpm and setting the rev-limiter at 6600 rpm. The results: 385 horsepower at 6000 rpm and 385 lb-ft. of torque at 4800 rpm. Both peaks occur at slightly higher rpm than the standard LS1. Low-end torque production was actually down slightly at lower engine speeds compared to the LS1, but the revised gearing in the M12 manual transmission made it a moot point.

As for the standard Corvette LS1, horsepower increased to 350 at 5600 rpm and torque went to 360 lb.-ft. at 4400 rpm for automatics (375 lb.-ft. for six-speeds). The F-bodies got a power bump, too: 310 horsepower and 340 lb.-ft. of torque for the base V8 cars, and 325 horsepower and 350 lb.-ft. of torque for the ram air models. The power increase for the 2001 LS1 was directly attributable to the integration of the LS6 manifold on both platforms. This was also the year that the 6-liter LQ4 truck engine finally received aluminum heads.

Power for the 2002 Z06 was upped to an incredible 405 horsepower and robust 400 lb.-ft. of torque. This was accomplished with the addition of a slightly more aggressive cam with more valve lift, improved catalytic converters, free-breathing airbox, low restriction mass air flow sensor, and lightened valvetrain with the use of lightweight hollow stem intake valves and sodium filled exhaust valves. Rumors abounded that the LS6 would find its way into a limited number of fourth generation F-bodies for 2002 to commemorate the cars' 35th anniversary and final year of production. It wasn't to be though, as the last year of F-body production was instead celebrated by cheesy vinyl decals on regular production F-cars. This was GM's chance to make a statement by showing the world it was serious about affordable performance for the masses. Unfortunately this never materialized, and GM let the Camaro and Firebird die a quiet death. That's truly a shame, for a limited run of LS6 F-bodies would have been the proper send off.

The LS6 engine is standard equipment for all Z06's.

The LQ9 engine makes 345 horsepower and 380 lb.-ft. of torque. Cadillac's Escalade is available with this 6.0-liter monster, making it one formidable SUV.

LQ9 Truck Engine

The LQ9 high-output 6.0-liter truck engine made its debut in all-wheel drive version of the 2002 Cadillac Escalade SUV. Based on the now-familiar Gen-III LQ4 iron block, the big news was the LQ9 cylinder heads. These aluminum castings have the same port design as the LS6 heads, but utilize a larger combustion chamber to keep compression at a pump gas-friendly 10:1. It is a truck, after all! Standard two-wheel drive Escalades received the 5.3-liter engine, while Escalade EXT and ESV both received the LQ9 engine as standard equipment.

Another golden opportunity was allowed to pass with the 2003 Corvette. Model year '03 was the platform's 50th Anniversary, and hopes were high that there would be something for the hardcore performance buffs to celebrate. Again, disappointment, as the Anniversary Edition was little more than a badge and vinyl stripe package. To add insult, no 50th Anniversary/Z06 combination was available.

The Future of the LS1

The Chevrolet SSR will begin production in 2003 as a high-performance compact truck equipped with Gen-III power. The SSR will feature GM's newest aluminum block 5.3-liter LM4 engine. The new variant is approximately 100 pounds lighter than the previous iron block, aluminum head version, and is expected to make around 285 horsepower. This aluminum version of the 5.3-liter is also slated for the some versions of GM's Trailblazer/Envoy line.

GTO Rebirth

Looming on the horizon is the rebirth of the Pontiac GTO. "The Great One" will return to U.S. highways in late 2003 when General Motors and its Australian subsidiary, Holden, begin production of a modern version of the legendary Pontiac GTO. Beginning with the 2004 model year, up to 18,000 GTO's will be produced annually at Holden's Australian production facility for sale in the United States.

The GTO will share the platform of the current Holden Monaro CV8 coupe, with some restyling to add Pontiac character. It is to be powered by the 5.7-liter LS1 V8, which will be

The 50th Anniversary Corvette was selected to pace the 2002 Indianapolis 500. Not surprisingly, no performance modifications were needed for it to perform its duties.

The death of the F-body in 2002 left a gaping hole for performance enthusiasts. The 2004 GTO should fill that gap very nicely.

mated to a choice of either a six-speed manual or four-speed automatic transmission. The rear-wheel drive coupe will likely be shod with 17- or 18-inch wheels and tires, and retain the Monaro's 2+2 bucket seating. Will it be a worthy successor to the legendary GTO of the sixties? Time will tell, but with LS1 power under the hood, chances are good.

While Juriga is the first to admit that nothing's ever perfect, the Gen-III engine is moving forward with small amounts of fine-tuning and tweaks. "Look for even more refinement for next generation as manufacturing processes and new materials allow us more flexibility in design. We are currently (spring 2002) working on the 2007 and 2008 engine programs. The individual coil ignition system lends itself well to Displacement on Demand and Parallel Hybrid fuel, GM's most recent economy initiatives."

What does the future hold for the LS1 and the Gen-III family of engines? Juriga was understandably tight-lipped, but offered: "There is a lot more potential that we have yet to tap. Look for us to unleash more power and torque in the years to come, while further improving both emissions and fuel economy."

When asked what the projected life span of Gen-III may be, Juriga quips, " I fully expect to retire before this engine's life is at an end, and I intend to be here another twenty years!" Now that you know where the LS1 came from, let's take a look at where it can go!

2 Bolt-On Performance Parts

LS1-powered racecars have become a common sight at the dragstrip. This new-age performer has humbled many a big-block owner.

Most people buy a performance car thinking it feels really powerful, and can't imagine how they'd ever improve it. But if you're like most gearheads, this thought has perished long before you get the car home. Once you've decided to take the leap and modify your car, you need to assess your objectives.

Do you want a low 12-second commuter with the reliability of an anvil, or a serious-as-a-heart-attack 9-second racecar? For most, something in between is just right. Once you decide on a goal for your project, time devoted to formulating a plan is well spent and will save you money in the long run. Having a detailed plan outlining your performance goals will prevent you from purchasing 12-second parts for a 9-second car or vice-versa. While that may seem like an obvious mistake, it is very common for enthusiasts to purchase a part only to have to replace it with a lighter/stronger/larger equivalent later as their car's performance continues to improve. It is far better to spend the extra money for superior components up front than to purchase and replace a part two or more times down the road. Buy parts that allow some performance headroom; you will grow into them before you know it! This chapter will look at the most common bolt-ons, those that give the most bang for the buck.

Why Bolt-Ons?

When it comes to modifying their performance car, bolt-on components are where most people begin and end. Bolt-ons generally refer to external components that can be easily changed or modified to improve the performance of the engine or the car as a whole. Expense associated with bolt-on components is usually fairly low and most of these performance parts are simple to install and do not interfere with emissions. In other words, these are do-it-yourself projects that can often be completed in your driveway on a Saturday afternoon. Some people consider power adders such as nitrous oxide or

A cotton gauze air filter like this Holley Power Shot is less restrictive than stock paper-type filters.

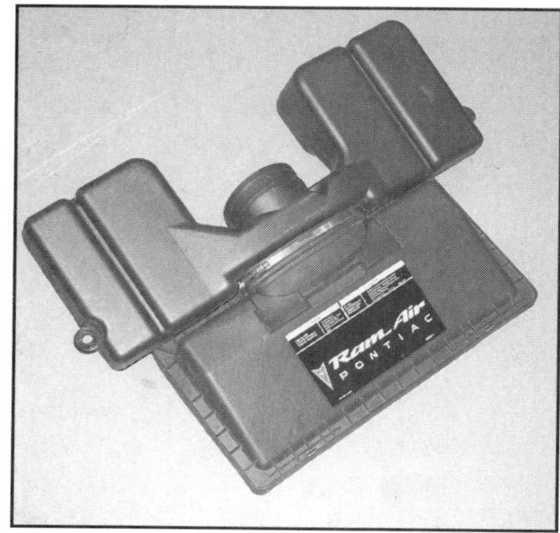

The stock F-body induction includes an air baffle behind the filter lid. This device was designed to muffle induction howl, which it does, but it creates power-robbing turbulence in the intake tract.

superchargers to be bolt-ons, but because they are a bit more involved, we will discuss them in their own chapter later on.

Cold Air Induction

If you think about the function of an induction system, its role is really quite simple. It is all about supplying clean, cool air to the engine. Because cool air holds more oxygen, it allows an engine to make more power than hot underhood air. Knowing this, you may wonder why GM stifles the LS1 with a restrictive induction system. The answer is that GM has more than just horsepower on its mind when designing a vehicle.

Manufacturers must strive to maximize the performance of their automobile without offending the delicate auditory senses and bank account of John Q. Customer. While it would be really cool to have an air-gulping factory-installed ram-air system on every new car, it is not economical on a mass scale to do so. But that is why there is a performance aftermarket, giving enthusiasts the choice to either install or build their own system.

Air Filters

Many people begin modifying their car with the installation of a high-flow air filter. Oil-treated cotton gauze filters such as those available from K&N will usually out-flow traditional paper filters and they can be cleaned, re-oiled and reused countless times. Holley and other companies offer comparable products.

While a high-flow filter is a good first modification, the filter itself is not an airflow restriction in a stock air intake system. Therefore, you should not expect much more than a couple horsepower. What you should do is address the induction system as a whole. Focus on minimizing restrictions, reducing induction air temperature, and ensuring a laminar flow of filtered air through the mass airflow sensor (MAF). Getting the air to the engine is a critical aspect of performance.

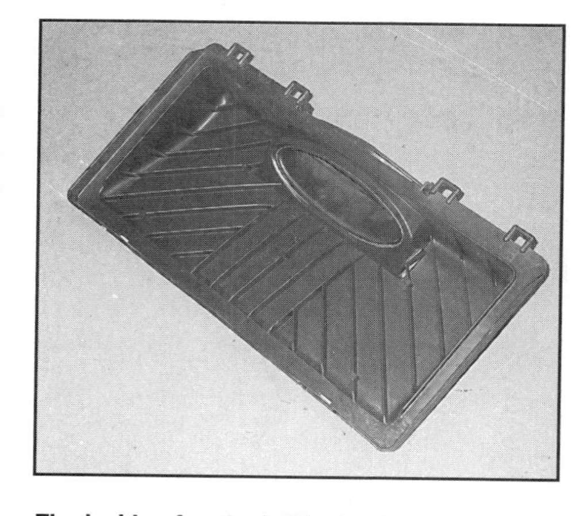

The inside of a stock F-body air lid is covered with noise-defeating ridges.

F-body Induction

Perhaps the best first upgrade for the F-body is the installation of an aftermarket airbox lid to go along with the high-flow filter. The stock airbox lid should be replaced because it is disruptive to smooth airflow. The interior roof of the stock lid is covered with a series of molded ridges designed to muffle the howl of induction. While they do an admirable job of this, they also disrupt airflow in the process. Located directly behind the airbox is a chambered baffle that further muffles the sound of rushing

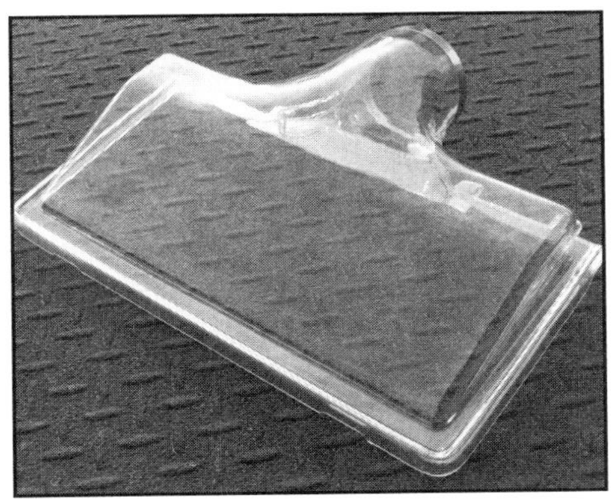

Motorsport Technologies, Inc. (MTI) was the first company to introduce a high-flow air lid for the F-bodies, and their piece is still in demand.

The LS1 Motorsports Direct-Flo lid incorporates all of the same features as the MTI lid, but sports a more curvaceous design.

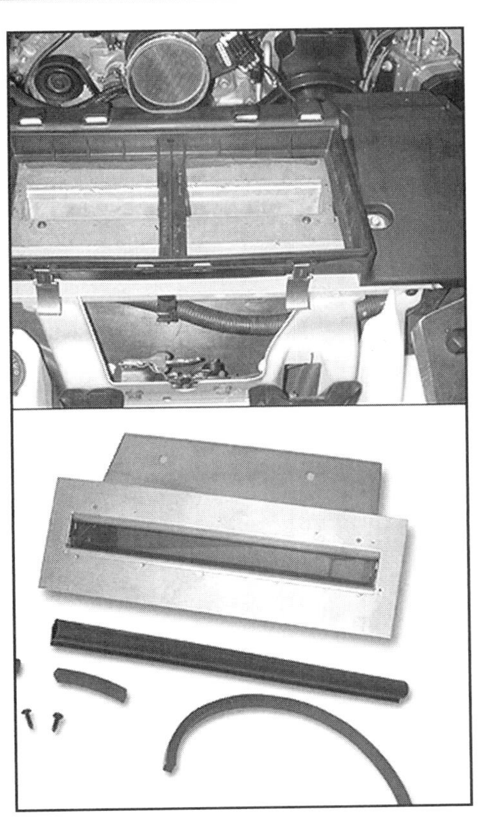

SLP's ram air system can free up a few horsepower by forcing ambient air into the intake.

Sun Coast Creation's ram air box improves induction airflow and transforms the look of an engine compartment.

air, but introduces turbulence to the air stream. An aftermarket air lid is an inexpensive upgrade that serves to eliminate both of these restrictions and will show a real power gain at the track or on the dyno. Lids are simple to install, and most are available with a K&N air filter. So how loud is it? Let's put it this way: As a performance enthusiast, you are highly unlikely to be offended by the sound of air rushing into your performance engine.

While there are many brands of airbox lids on the market, virtually all of them look and perform about the same, so you really can't go wrong. Most lids are direct-fit replacements for the factory unit. They eliminate the intake sound baffles and feature a provision for the Intake Air Temperature (IAT) sensor. Most engines gain in the neighborhood of 10–12 rwhp from this modification.

Cold Air Systems

An excellent companion to a high-flow lid is an off-the-shelf ram air system such as the one offered by Street Legal Performance (SLP). These are designed to force cold, high-pressure air into the intake tract. Because this type of product depends upon the vehicle moving down the road, dyno numbers don't show much, if any, gain. Track testing usually shows 1/4-mile improvements of about two-tenths of a second and 2 mph.

Free Cold Air Mod for the F-body

There is a no-buck alternative to the

The "Free Cold Air Mod"

The first step is to get the stock lid out of the way. Gently pry up the two black plastic retaining clips on either side of the air silencer, and then loosen the clamp that fastens the intake bellows to the throttle body.

Remember to unplug the mass airflow and intake air temperature harnesses before removing the assembly.

With the lid and bellow out of the way, you can now remove the air filter, tray and the four bolts retaining the airbox base to the core support. Remove the base and set it aside for the time being.

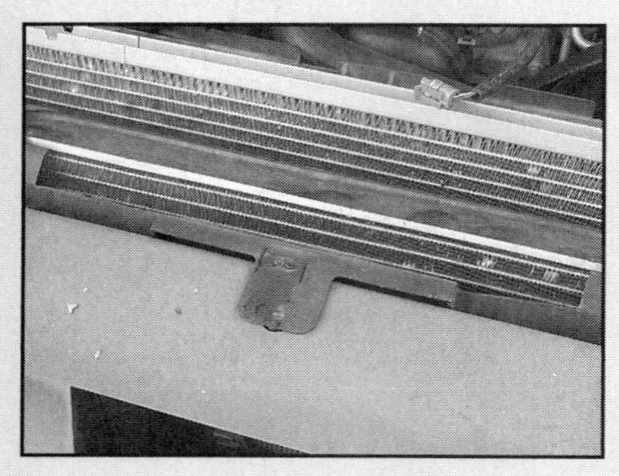

This narrow opening is all that supplies air to the LS1. Now you can cut out this flap as shown.

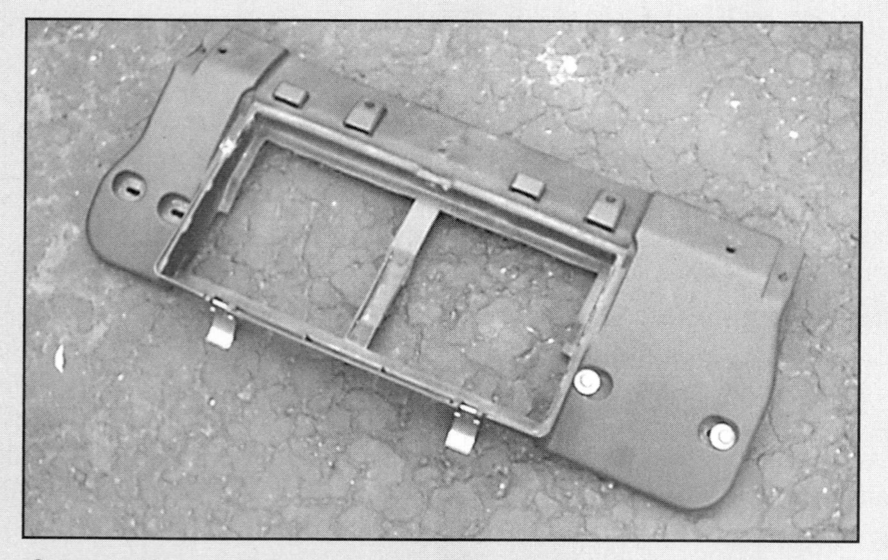

Cut away this part of the airbox base to match the flap above. A Dremel or even a fine-toothed hacksaw blade will do the trick. Simply reverse the above steps to reassemble the intake tract. If you are adding an aftermarket lid, you will not reinstall the air silencer.

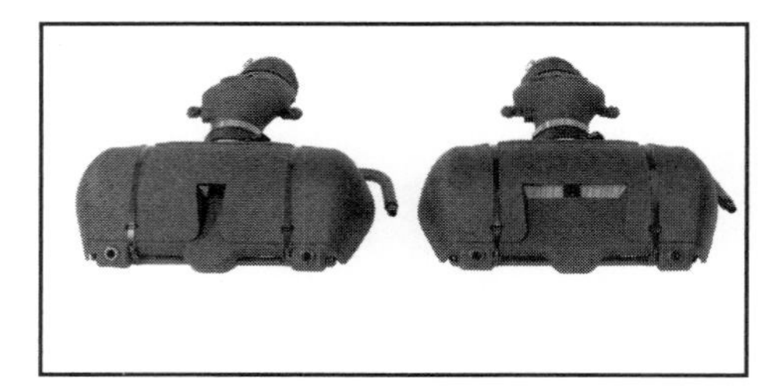

Compared to the 2001 component (L), the 2002 Z06 stock airbox assembly (R) features an enlarged inlet area to feed the 405 horsepower LS6.

commercially available ram air systems: the "Free Cold Air Modification." The F-body (non-ram air cars) draws its air from beneath the car by way of a cavity in front of the air-conditioning condenser. The downward-facing airbox inlet is a narrow opening that is restrictive to airflow. This is easy to fix with a little creative cutting of the plastic inlet, and the bottom of the airbox. See the sidebar for step-by-step instructions on how to complete this modification. The Free Cold Air Mod is of questionable worth on a ram air car, as the tall air lid base is fed air from the front by way of the hood scoop. You are likely to gain just as much by removing the water baffles from the car's hood scoop.

Another option is an air box designed for aftermarket ram air hoods. Sun Coast Creations has a box for use with their Z-style or Raptor hoods that features a huge upward facing K&N filter for minimal restriction. As with the stock ram air hoods, you might consider removal of the water baffles from the hood scoops for maximum performance.

Corvette Induction

Although not as restrictive as those found on the F-body, the air intake tract on the Corvette can still be improved upon. The stock airbox and

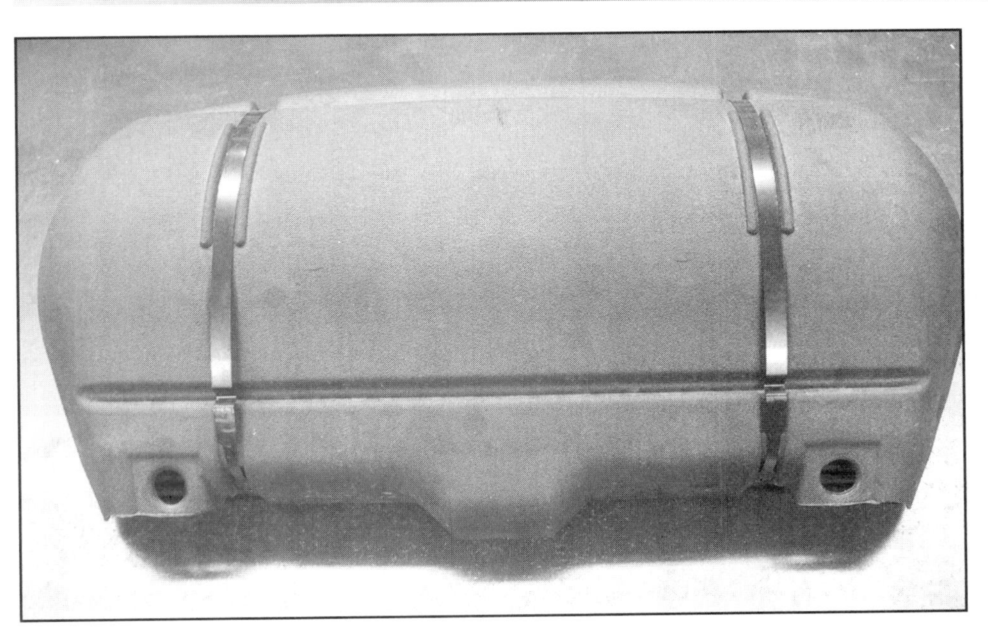

A stock C5 air box cover can be easily modified for improved airflow. A sharp utility knife, plus a bit of patience is all you'll need to do the job.

Far and away the most popular front breather style cold air intake system for the C5 is the Donaldson Blackwing. It is a wedge-shaped, oil-impregnated cotton gauze filter that completely replaces the stock airbox assembly. This is arguably the best bang-for-the-buck C5 cold air induction system on the market.

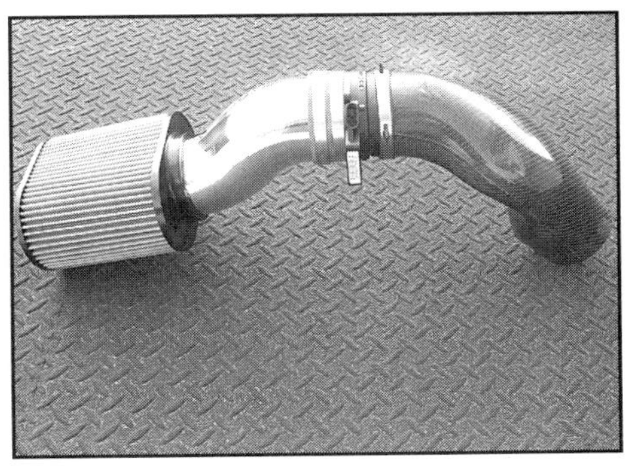

The Halltech TRIC/MAG combination may be the most elegant of the bottom breathers, utilizing titanium and carbon fiber construction and a proprietary cotton gauze air filter.

"bridge" went pretty much unchanged from 1997 through model year 2000. The 2001 Z06 had an improved airbox with an enlarged inlet, and the 2002 version's inlet was opened even further. This means that you will not see quite as noticeable results from cold air systems installed on these cars, since the have less restrictive induction systems to begin with.

There are two main styles of aftermarket cold air induction systems available for the C5: "front breathers" and "bottom breathers." Each have their advantages; it is up to you to decide which best fills your needs.

Front Breathers—"Front breather" is slang for aftermarket induction systems that, as implied, draw their air from the front of the car, rather than the bottom ("bottom breathers"). Front breathers draw air from the same area as the stock airbox, and require no modifications to the car. Installation typically takes no more than a few minutes, and can be accomplished by anyone familiar with the Lefty-Loosey Principle. The downside of the front breathers is the lack of a direct source of outside air. The exception is the Z06, which has air inlets in the front bumper cover.

Bottom Breathers—As the name implies, bottom breathers draw their air from the underside of the car, usually through a hole cut in the radiator shroud. Bottom breathers have a slight edge in performance over the front breathers, but usually no more than 3 or 4 horsepower. This margin is practically nonexistent in the case of a Z06, as a result of their front-mounted air inlets. When using a bottom-breather, extra caution must be exercised in heavy rain or when passing through puddled water, as it is not unheard of for intakes of this type to ingest water. GM is not likely to warranty a hydro-locked engine caused by an aftermarket induction system!

It's a good idea to remove the radiator shroud from the car in order to modify it for the installation of a bottom breather. Most kits include a template, and the plastic is easy to cut with a Dremel or a utility knife.

A stock F-body LS1 mass airflow sensor has a bore of 74mm and utilizes a honeycomb matrix to smooth airflow past the sensing elements.

Mass Airflow Sensors

Once through the air filter assembly, the next component encountered by incoming air is the mass airflow sensor (MAF). The MAF is an information sensor that measures the quantity of air flowing into most sequentially fuel-injected engines, including the LS1. The MAF allows the engine management system to accurately calculate the air/fuel ratio to optimize engine performance and fuel economy.

MAF sensors use an electronic hot wire sensor to measure airflow. The sensing element provides an output voltage reading that is directly proportional to the amount of air that is entering the engine. The advantage to this method is that it reduces the calculations required to control fuel flow and ignition spark advance by the power-train control module (PCM). The MAF can react with speed and accuracy to sudden changes in airflow, which improves overall vehicle performance and drivability. The LS1 MAF utilizes a die cast aluminum body with a 74 mm bore size, while the LS6 MAF has a plastic body with an 85 mm bore size and an integrated IAT sensor. The 85 mm sensor is also found on 6-liter truck engines.

Porting MAFs

A very common modification is the porting of the stock MAF. Engines equipped with an aluminum-bodied MAF can get a small power gain by smoothing the inside of the housing and removing parting lines left over from the casting process. Partial or complete removal of the central airfoil can also eliminate a substantial flow restriction. An hour or so with a die grinder will generally whip the MAF into shape. If this seems like too much work, most vendors sell billet aluminum or plastic MAF ends that are an easy bolt-on replacement.

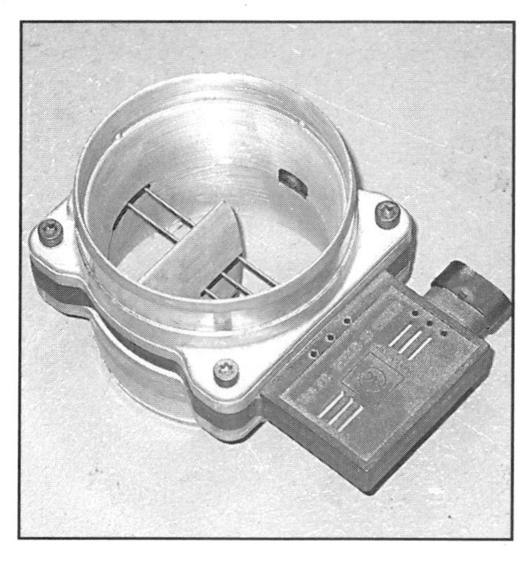

Here is a look at a ported F-body MAF. Notice that the bore has been smoothed, and that the screen and half of the dividing airfoil have been removed. All of this is done in an effort to increase airflow.

MAF Screen—The mass airflow sensor has a honeycomb screen at the entrance into the sensor body. The purpose of this screen is to create a smooth, laminar airflow upon its entry to the MAF, which allows the unit to obtain a more accurate reading. It also has the undesirable effect of limiting the airflow through the unit. It is common practice to remove this screen. Some cars react with a gain, but some react with poor idle and tip-in stumble. If you are modifying your

Billet aluminum MAF ends are simple to swap on, and saves you the hassle of porting the stock parts.

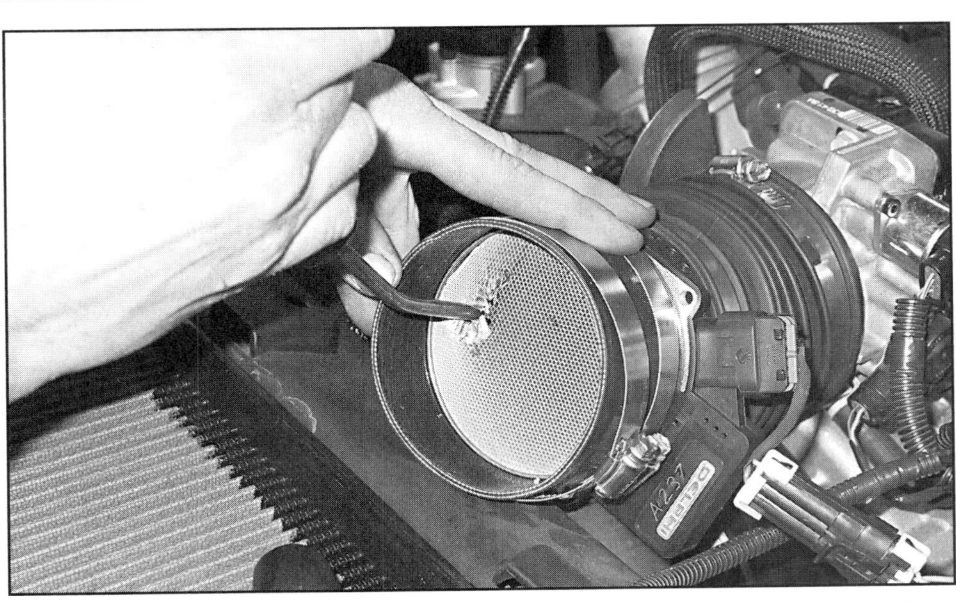

Mass airflow sensor screen removal is most easily accomplished with a pick as shown here. Be forewarned though: Using this method, you will not be able to reinstall the screen.

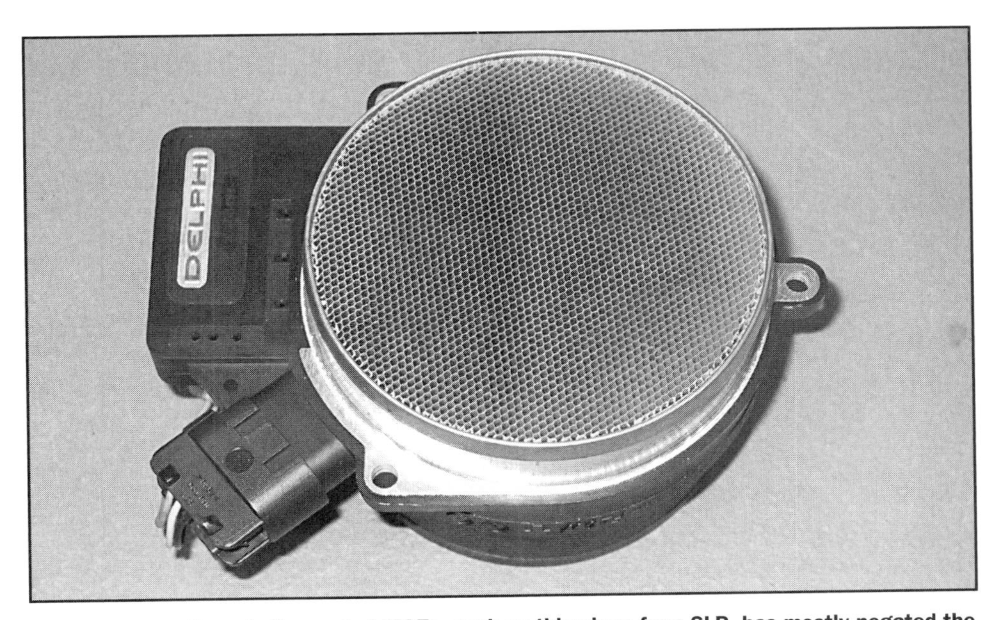

The new generation of aftermarket MAFs, such as this piece from SLP, has mostly negated the need to port a stock MAF.

car for all-out power, go ahead and rip it out. Otherwise, you may be better off leaving the screen intact. GM has deleted the screen starting with the 2002 LS6.

Ported MAFs & Ignition Timing—Porting the MAF not only smoothes induction airflow but also instigates an increase in ignition timing. This occurs because a ported MAF will report less airflow at identical engine RPM and manifold pressure as an unported sensor. This results in a lower calculated load causing the PCM to utilize a look-up table for a different curve.

For example, the PCM may command 28° timing advance for 100% calculated load at 6000 rpm. But at 78% calculated load at the same rpm, the PCM map may require 29-30°. Essentially, the PCM perceives that the engine is achieving 6000 rpm under a lighter load condition and adds ignition timing. This is something you should observe using a scan tool. Be certain that the fuel in the tank has sufficient octane to support the additional ignition timing, otherwise detonation may occur. More ignition timing, up to about 28° for a lightly modified LS1 can result in a toque increase.

Aftermarket MAFS—The early generations of aftermarket "recalibrated" MAFs gained a bad reputation, and rightly so. Many were little more than plastic ends bolted to an otherwise stock MAF, accompanied by some outrageous power claims. Equal (and oftentimes better) results were had by simply porting the stock ends or swapping them out for billet ends.

Times have changed, though, and the latest sensors can deliver a real performance gain to an engine that needs additional airflow. SLP Performance Parts offers an 85 mm MAF calibrated specifically for use with the earlier LS1s. The SLP piece comes with an adapter harness, so it is completely plug and play.

Pace Performance Parts offers their line of "Velocity Volume" recalibrated 85mm MAFs. These MAFs can be ordered pre-calibrated for several

Unlike previous designs, the thermostat on all Gen-III engines is located on the water pump housing, shown here.

Cat-back exhaust systems like this one from SLP will net a small power increase and a healthy rumble.

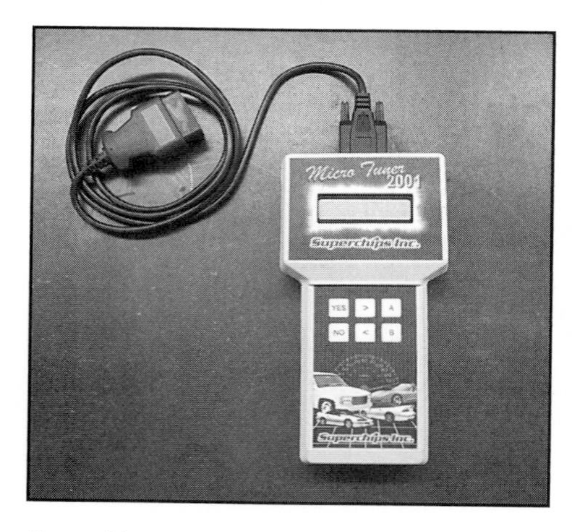

Superchips offers its handheld programmers with standard and custom tuning options.

different injector sizes, which allow you to upgrade injectors without the expense of a PCM re-flash. For more on fuel injectors, see chapter 9.

Other Modifications

Cooler Thermostat

While nearly all of the performance vendors offer cooler thermostats, they are rather costly. The stock piece is remarkably simple to modify, and can be quickly and easily adjusted to open and close at just about any temperature between 160° and 195°. See chapter 4 for an easy step-by-step tutorial.

Exhaust Systems

High-flow exhaust systems are an effective way to increase engine performance on most cars. Cat-back exhaust systems, so named because they generally replace all components aft of the catalytic converters, are a convenient "exhaust system in a box." In the case of LS1-powered cars, however, this is not an area to expect huge power gains. GM did an outstanding job designing these exhaust systems to flow well with minimal noise. Typical gains from the installation of a cat-back are in the neighborhood of 10 rwhp. Cat-back exhaust systems are explored more thoroughly in chapter 8, as are headers and X/Y - pipes.

Handheld PCM Programmers

Handheld programmers are a popular items that work well for adjusting the speedometer for rearend gears, and modifying the temperature at which the cooling fan is activated. While most offer an off-the-shelf performance tune program, these are typically ineffective and it is not unheard of for these programs to actually decrease power. We'll take a closer look at program boxes in chapter 10.

Underdrive Pulleys

Underdrive pulleys lessen parasitic losses by slowing the drive speed of accessories such as the water pump, alternator and power steering. Companies such as ASP and MTI offer smaller diameter crank pulleys and larger diameter alternator pulleys to accomplish this. These are an easy bolt-on that will result in a power increase in the neighborhood of up to 5 rear-wheel horsepower.

Short Belt—The short belt mod replaces the serpentine belt with a shorter version bypassing the power steering pump. Get a 51.5-inch, 6-rib serpentine belt, and route as normal,

MTI and ASP are two companies offering underdrive pulleys for the LS1.

bypassing the power steering pulley. You may need to remove the lower alternator bolt, as it is often an obstruction. Because you lose all power assist this mod is recommended for drag strip only cars. You can expect to gain 3–4 horses. You might also want remove the belt for the air-conditioning compressor at the track.

Intake Manifolds

When you open the hood of an LS1-powered vehicle, what's the first thing you notice? Chances are, it's the unusual plastic intake manifold. Though you might have expected to see an aluminum intake on a performance engine, plastic intake manifolds make a lot of sense: They are easier to mold into complicated shapes, cheaper to manufacture, weigh less, and run cooler than their aluminum counterparts. For those reasons, you can see why plastic intake manifolds are becoming standard in today's OEM cars and in the aftermarket in the not-so-distant future.

Technical Overview

Plastic may be the generic name, but the specific material is Dupont Zytel, Nylon 6.6, which is Nylon with glass fiber reinforcement. It is phenolic in nature, which means it

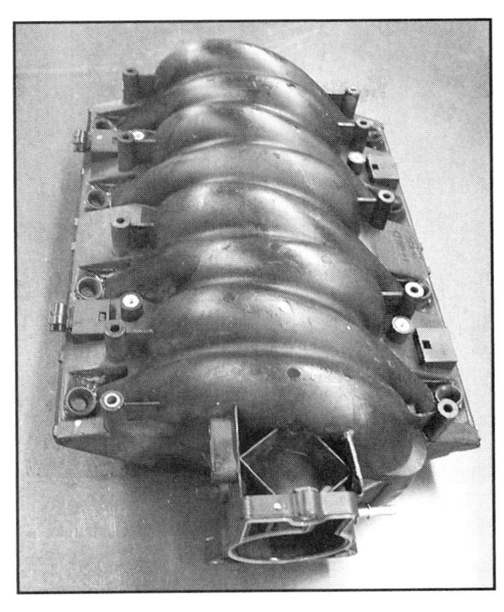
The nylon construction of all Gen-III intake manifolds represents a growing trend in the auto industry. While rather unorthodox in looks, the material has proven to be cheaper, more pliable, and more thermally efficient.

doesn't readily transfer engine heat to the incoming charge air. This also serves to thermally isolate the fuel rail and injectors mounted on the manifold. Reduced heating of the fuel system translates to improved fuel control during extreme temperature operation. The manifold is also quite lightweight, tipping the scales at just 8 pounds.

In order to minimize overall engine height, the plenum occupies the space in the valley beneath the runners. This design allows smooth, radial entry of the incoming air to 260 mm runners. These runners route air up and back over the plenum into each intake port at the cylinder heads. In order to increase flow velocity into the port, the runner cross-sectional area decreases with length traversed. The long runners and relatively small camshaft serve to minimize intake charge reversion. This results in a small low-rpm torque increase with only a minor sacrifice of high-rpm horsepower.

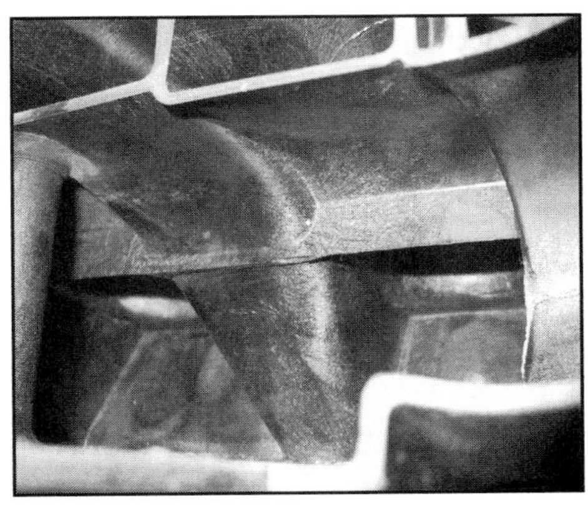
As this cutaway of an LS6 intake manifold shows, the Gen-III engine situates the plenum beneath the individual runners. Air is drawn upward from the plenum to the runners.

Referring again to our cutaway model, you can see the precise fuel injector location. It positions the injector out of the airflow, and directs its spray directly at the intake valve.

The throttle body is sealed to the intake manifold by a one-piece, pressed-in-place silicone gasket. The Manifold Absolute Pressure (MAP) sensor is located at the rear of the manifold and sealed by an O-ring. The LS1 intake handles only air; there is no water crossover, and no oil contact due to the sealed valley cover.

The fuel rail assembly and its eight fuel injectors are fastened to the intake by four bolts. The injectors are sealed in their individual manifold bores with O-rings. The fuel injectors are targeted at the backside of the

GM Manifold Specifications

Intake Manifold	GM Part Number	Plenum Volume	Runner Length	Runner Volume
1997–2000 LS1	12556333	5.06 liters	262 mm	.536 liters
2001–2002 LS1/LS6	12561182	5.19 liters	262 mm	.541 liters
2002+ LS1/LS6	12573572	5.19 liters	262 mm	.541 liters
1999–2001 LQ4	25321788	4.0 liters	263 mm	.513 liters
2002 LR4/LM7/LQ4	12573944	4.0 liters	263 mm	.513 liters
2003 LR4/LM7/LQ4/LQ9	12574009	4.0 liters	263 mm	.513 liters
2002 LQ9	25316654	4.0 liters	263 mm	.513 liters
1999–2001 LR4	25321788	4.0 liters	263 mm	.513 liters
1999–2001 LM7	25321788	4.0 liters	263 mm	.513 liters
2003 LR4	12574009	4.0 liters	263 mm	.513 liters

Head & Manifold Flow Comparison: Airflow in Cubic Feet per Minute

Cylinder Head Head	Intake Manifold	Inches of Valve Lift & CFM .200"	.300"	.350"	.400"	.450"	.500"	.550"	.600"
Stock LS1	LS1	136	184	200	214	222	227	229	235
Stock LS1	LS6	136	186	206	223	227	236	241	242
Stock LS6	LS1	156	199	212	224	232	238	243	247
Stock LS6	LS6	154	204	220	235	247	257	263	265

Here is a close look at the intake manifold cylinder head port gasket area with its captive gasket. Note the injector boss at the top of the port.

intake valves, promoting better fuel atomization and reduced fuel puddling, which in turn increases power and improves fuel efficiency.

Eight captive silicone gaskets are pressed into grooves in the manifold mounting flanges, sealing the manifold to the cylinder heads. The end gap seals are open-cell foam, to minimize noise radiated from the manifold to the knock sensors located in the valley cover shield. The intake manifold, fuel system, throttle body, and other hardware are pre-assembled into a fully Integrated Air and Fuel Module (IAFM). The IAFM is fastened to the cylinder heads with ten screws and is flow and leak checked prior to delivery to the engine assembly plant.

The Myth of the Restrictive LS1 Manifold—Immediately upon its introduction in 1997, the LS1 intake manifold was labeled very restrictive by countless self-proclaimed "experts." This is more urban legend than fact, however, as extensive testing has shown that the LS1 manifold will comfortably support 400 rear-wheel horsepower. It has its limitations of course, as the LS1's plenum volume becomes a limiting factor in high-rpm or large displacement applications. If your engine falls into one (or both) of these categories, upgrading to the LS6 manifold is definitely a good plan. But you're in luck if your car is a 2001 or later, as this manifold became standard equipment in that year.

LS6 Manifold—Calling the 2001-and-up intake the "LS6 manifold" is actually a bit of a misnomer (but it does minimize confusion). The LS6

A typical 1998–2000 F-body intake manifold with its attendant EGR hardware.

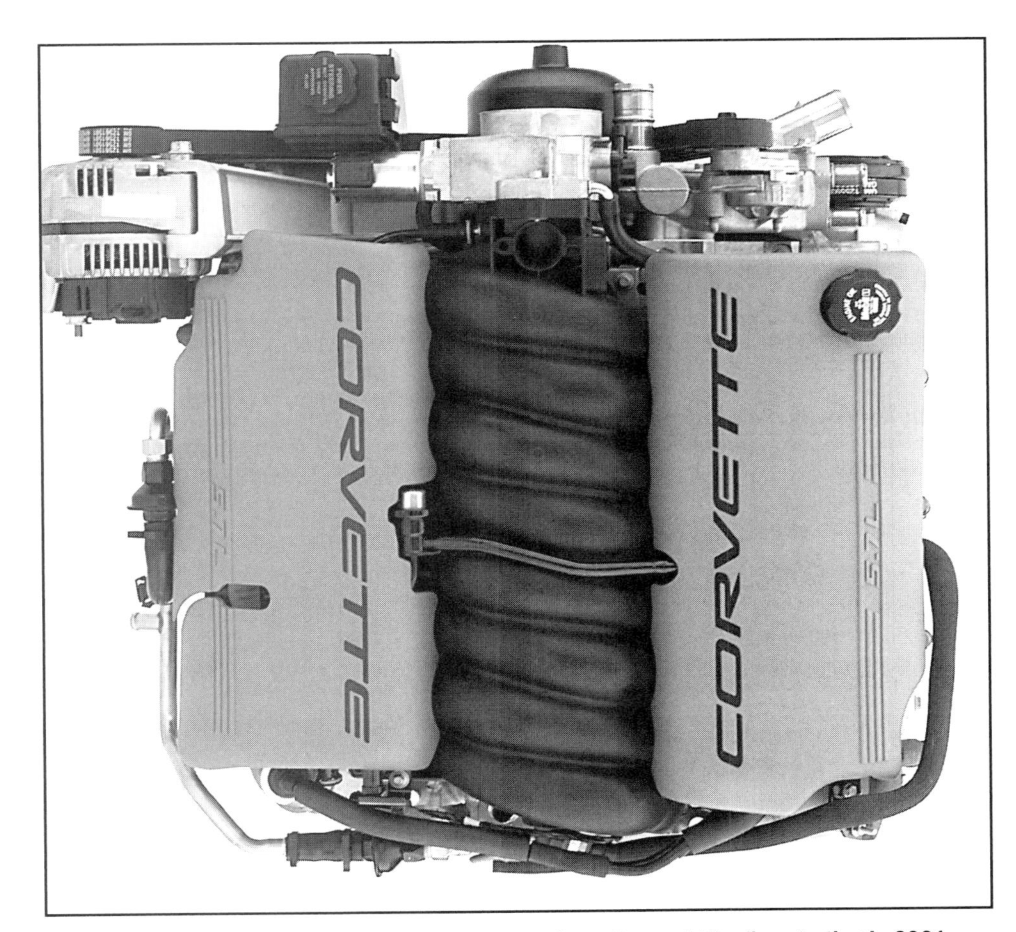

The LS6 intake manifold is standard equipment for Corvettes and F-bodies starting in 2001.

manifold is so dubbed because it was introduced as original equipment on the LS6 engine in the 2001 Corvette Z06. But it also became standard equipment on all LS1 cars that year, adding 5 horsepower in the process. Notable features include an increase in plenum volume, revised runner lengths (see table above) and elimination of the EGR system. The LS6 manifold is molded from the same Nylon 6.6 as the LS1 piece and, besides the missing EGR provision, is nearly indistinguishable once installed.

When added to an aggressive head and cam equipped engine, the LS6 manifold has shown up to a 20 rear-wheel horsepower gain. Many people assume this manifold is worth an automatic 20 horsepower on any engine, but this is just not the case. The truth is that the manifold by itself is worth more like 7 or 8 rear-wheel horsepower and a similar gain in torque on a mostly stock engine. Until such time as the intake is unable to supply sufficient airflow to the engine and becomes a bottleneck, there is not a lot to be gained by replacing it with a more efficient piece. This generally occurs around the 400 rear-wheel horsepower level. Most cars with ported cylinder heads and more aggressive cam will realize a gain from the LS6 manifold. Those with a stroker or power adder will definitely benefit from an intake swap.

If you decide that the LS6 manifold is right for your engine, you're in luck because the LS1 and LS6 manifolds are interchangeable. This is a great upgrade and quite easy to do, but you will have to make a few simple changes to properly complete the swap. See the sidebar for more details on performing this installation.

How to Install an LS6 Manifold

Installation of the LS6 manifold is a very straightforward procedure on either an F-body or a Corvette, but here are a few tips to help make it painless as possible.

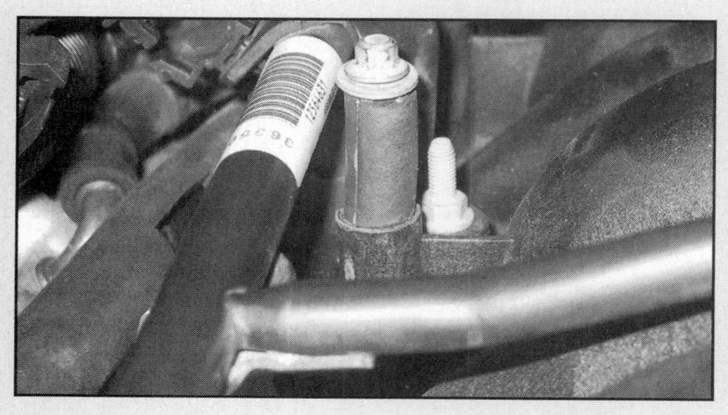

The trick to getting the intake off an F-body is to keep the rear bolts out of the way. Because of their proximity to that pesky cowl, they cannot be removed with the intake in place. Slip a short length of split fuel hose over the bolt to hold it in place. This allows easy manifold extraction. No such effort is required on Corvettes.

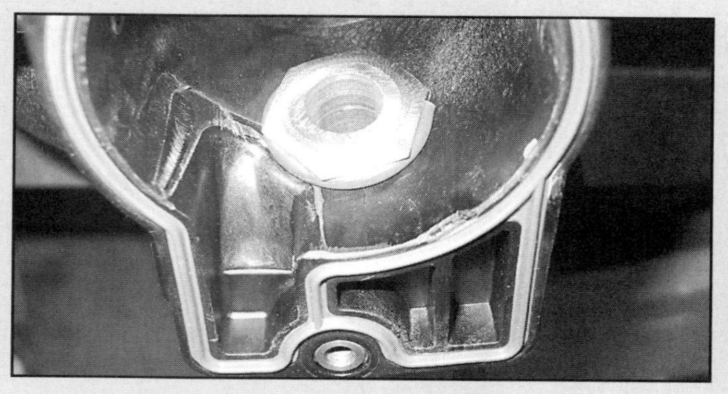

The LS6 manifold does not have an EGR provision, so you will have to decide if you wish to retain the EGR system. If you don't have to contend with the smog Nazis, you might choose to eliminate the EGR from your car. If you do so, you will have to fabricate a small block-off plate for the passenger side header or exhaust manifold. You can then remove the EGR valve from the car, but you will have a Service Engine Soon light to contend with. You can either live with the SES light, or have a custom PCM programmer disable the EGR functionality.

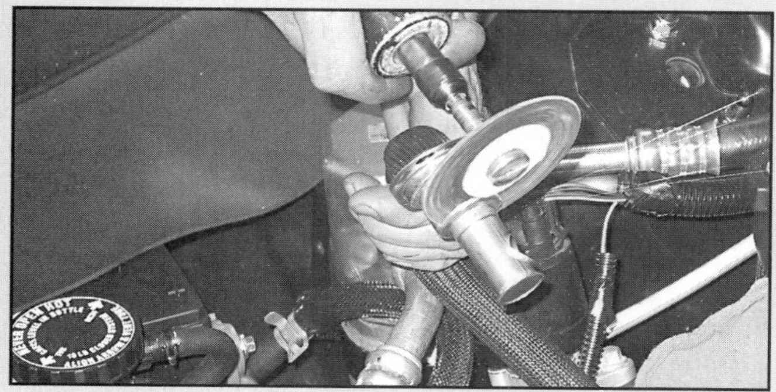

While GM does not equip the LS6 manifold with an EGR provision, nearly all the aftermarket performance shops do! For those that wish to retain EGR functionality, Agostino Racing, Lingenfelter, SLP, and various other vendors modify the manifold to accept the EGR feed. Shown here is an SLP-modified intake complete with a brass compression fitting. In order to insert the EGR tube into this ferrule, you will have to cut off the lower portion of the tube as shown here.

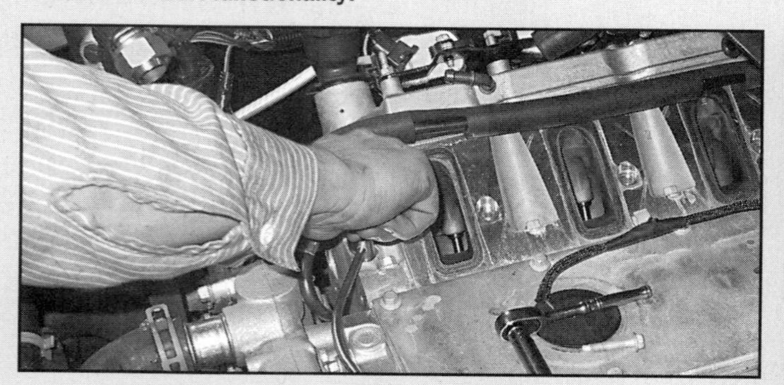

Because the increased plenum volume leaves precious little room in the intake valley, the under-manifold coolant lines need to be re-routed. You should replace the coolant lines with those from a '01 or newer engine, which are available from GM. They'll swap right on and are cheap to boot.

After installation, the new LS6 intake manifold is nearly indistinguishable from the original LS1 unit.

The 6-liter truck manifold is not a thing of beauty. Many mistakenly believe it to offer a performance gain over the LS1 manifold, but testing has disproved this theory, as these manifolds are designed for low end torque production.

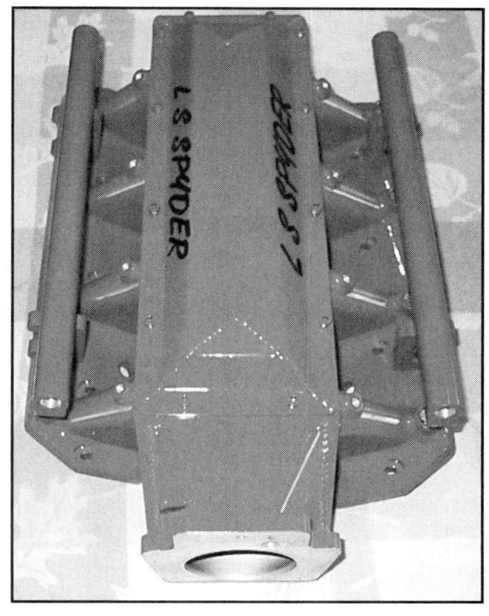

Nitrous Warehouse's Spyder Sheet metal manifold is a thing of beauty, but testing has shown it best suited to race-only applications.

Holley's cast aluminum intake manifold closely resembles a production piece in appearance and performance. Its main advantage is that it can be ported to allow additional airflow.

Typical Ported Heads and Intake Combinations

Cylinder Head	Intake Manifold	Horsepower	Torque
Ported LS1	LS1	406	393
Ported LS1	LS6	423	405
Ported LS6	LS1	413	400
Ported LS6	LS6	435	416

Example is 1998 Camaro six-speed with stock shortblock, Lunati 222/230° camshaft, 1.75-inch long tube headers, off-road Y-pipe, B&B cat-back exhaust. Heads were ported in the same manner with 2.055/1.60 valves installed.

Truck Manifolds

The intake manifolds offered on the 6.0-liter truck engines feature slightly longer runners but less plenum volume than LS1 car manifolds. Though they are a direct bolt-on to any Gen-III engine, cowl and clearance issues make installation a problem in car applications. And since their forte is low-end torque rather than top-end power, the LS6 manifold is a much better upgrade for most hot rodders.

Aftermarket Manifolds

The inherent efficiency of the factory LS1 and LS6 manifolds have left the aftermarket puzzling over how to build a better mousetrap. Predictably, this has led to two types of aftermarket intakes: sheet metal and cast aluminum. To date, there are no plastic manifolds available from the aftermarket.

Sheet Metal Manifolds—A handful of companies such as Nitrous Warehouse have produced sheet metal intake manifolds. Most have resembled the large plenum/short runner configuration of the Gen-II (LT1) manifold. On street engines these types of intakes commonly show losses of 30 to 50 lb.-ft. of torque through the mid-range. Horsepower is down too, but makes a

All Gen-III engines utilize a large single-bore throttle body, such as this LS6 piece beneath the hood of a Corvette Z06.

comeback before breaking even around 6200 rpm and continues to gain steadily from there. This makes for a peaky engine with poor throttle response in normal driving conditions. Because of these shortcomings, sheet metal manifolds are best used as race-only pieces. They have their place, but that place is not under the hood of a street car.

Cast Aluminum Manifolds— Holley is producing a cast-aluminum replica of the LS6 manifold designed by noted performance engine builder John Lingenfelter. The cast aluminum design is said to be more durable than the factory composite pieces, but the tradeoff is more transfer of engine heat to the incoming charge and some extra weight. The runner floors and plenum area contain extra material to allow porting. The bottom of this manifold is removable to facilitate this. The aluminum manifold's thicker walls can be easily machined for the installation of direct port nitrous fogger nozzles.

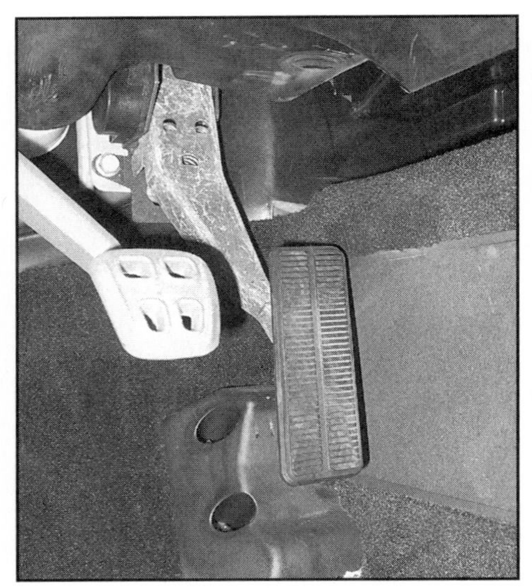

Drive-by-wire means that there is no mechanical connection between this Corvette throttle pedal and the throttle body on the engine.

Throttle Bodies

The F-body is equipped with a conventional cable-actuated throttle control system. But if you've taken anything more than a cursory glance under the hood of a C5, you may have noticed something unusual about the throttle body: no throttle cable! Mounted to the side of the throttle body is an Electronic Throttle Control (ETC). The LS1 in the 'Vette is the first drive-by-wire application in a GM car and ETC has subsequently been added to Cadillac's LQ9 equipped trucks.

Drive by Wire

The ETC integrates the functions of the mechanical throttle, Idle Air Control motor (IAC), cruise control module and the traction control throttle relaxer. System architecture provides a dedicated serial data link between three ETC components: Accelerator Pedal Module (APM), Throttle Actuator Control Module (TACM) and the brains of the outfit, the Powertrain Control Module (PCM).

A ported stock throttle body will supply sufficient air to all but the most raucous LS1. Note the smoothed walls, which encourage airflow.

The APM consists of a stamped metal assembly for the housing and lever. It utilizes compression springs and friction disks to provide the desired Force v. Displacement. This allows drivability to be tuned in a manner consistent with vehicle character. An integral three-track position sensor provides driver intent sensing.

The Throttle Actuator Control Body is die-cast aluminum with a 75 mm bore. A DC brush motor controls throttle blade position through two-stage gear reduction. A two-track throttle position sensor provides blade position feedback to the TACM. The TACM reads and serially communicates the actual pedal and throttle blade positions to the PCM, which then communicates the desired blade position back to the TACM.

Ported Throttle Body

Throttle body porting is an easy procedure that can be done by nearly anyone with a die grinder or Dremel. Though the casting is too thin to be bored to accept a larger throttle plate, the stock throttle body will adequately supply all but the most aggressive

BBK offers a 80mm throttle body for the LS1. If you are using a production manifold, stick with a ported stock throttle body.

LS1. Only when joined with a very large displacement or forced induction engine would a larger throttle body be beneficial. After removing the throttle body from the manifold, simply clean up the casting flash and bell-mouth the inlet to promote smooth airflow past the throttle blade. This modification performed alone on an otherwise stock engine will generally produce 2 to 3 rear wheel horsepower.

Aftermarket Throttle Bodies

Aftermarket throttle bodies offer a larger bore (usually 80 mm) and throttle plate. Gains over a well-ported stock throttle body are minimal when used on a GM manifold because the manifold inlet cannot be enlarged to match the larger bore of the throttle body. Unless you intend to move to an aftermarket manifold with a larger inlet bore, a larger throttle body is unlikely to show any gain over a ported stocker.

Free Mods

The following sidebars will instruct you on how to make some subtle improvements to your LS1. Subtle, because any one change by itself is unlikely to produce a notable difference, but when you add all of the mods together, you will find their cumulative effect worth the effort.

Once you have everything buttoned up, you should start the engine and check for vacuum leaks, coolant leaks and SES lights.

EGR Modification

Remember the EGR fitting protruding from behind the throttle plate, shown in the last photo of the WOT sidebar? This airflow obstruction is easily removed without affecting EGR functionality. Note that this modification is intended only for 1998–2000 F-bodies. All C5's and 2001 and newer F-bodies are equipped with induction systems without EGRs.

The EGR fitting is secured to the manifold with a single bolt. Once you have the fitting exposed, use a die-grinder or Dremel with a cut-off wheel to remove 1.25-inches from the end of the fitting as shown here. No Dremel? A hacksaw will work fine, but be careful not to over-stress the EGR plumbing; it becomes brittle from prolonged exposure to hot exhaust gases.

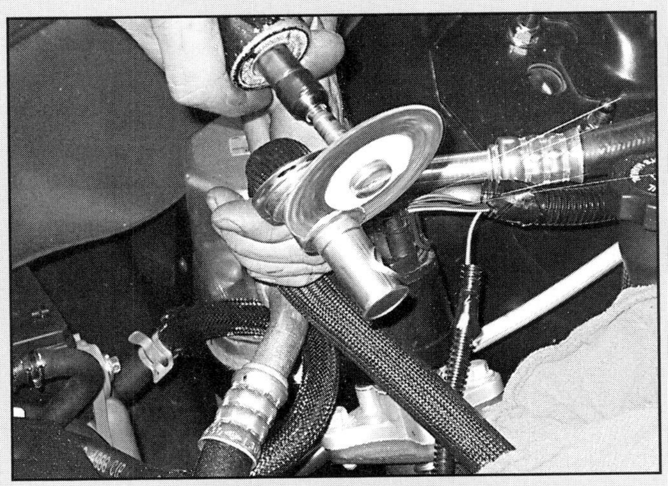

When you reinstall this piece, be careful to properly seat the O-ring. If it is not correctly located, you will have a massive vacuum leak and an erratic idle. Do not overtighten the single bolt fastening the hold-down plate to the intake manifold. The threaded brass insert on the manifold is very easy to dislodge, and doing so is very likely to ruin your day.

How to Modify the Throttle Stop for 100% WOT

The LS1's throttle-blade on many F-bodies does not completely open when the accelerator is mashed. By carefully grinding away a small portion of the wide-open throttle stop, you can allow the throttle blade to rotate to a perfect 90° angle that will allow maximum airflow into your engine. Because it is equipped with Electronic Throttle Control, no such problem appears to exist on the Corvette.

The fix for this problem is to remove some material from the WOT stop. As you can see in this photo, only a small amount of grinding is necessary. This is easily accomplished with a Dremel or small die-grinder, though a small file would do in a pinch. The procedure is best performed on the car, as you want to take only as much material as necessary. Working with it on the car allows you to easily check your progress with a scan tool. Before you start grinding, you should stuff a rag in the throttle bore to keep any wayward chips out of your engine.

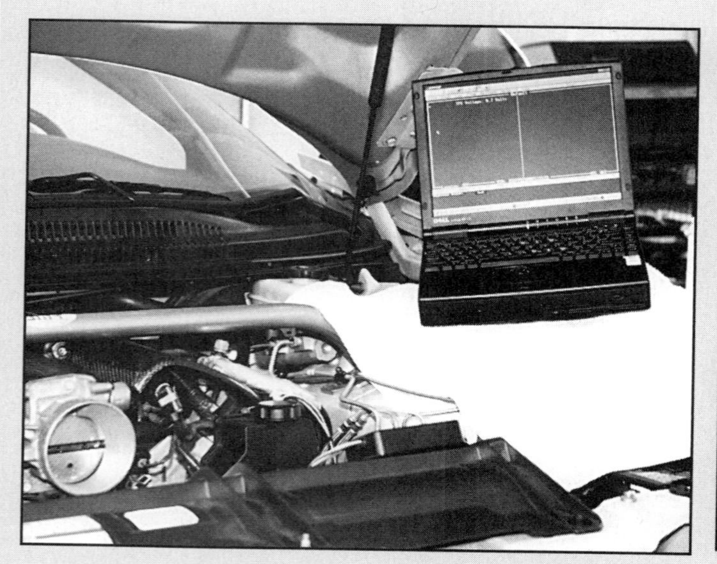

The best way to check your progress is with a scan tool that will display throttle position sensor voltage. Some, but not all versions of Auto-Tap will display voltage while others show only a percentage number or throttle angle. You could also back-probe the throttle position sensor and monitor the voltage with a digital volt-ohm meter, but the scan tool is a better approach because you will be able to monitor the voltage as read by the PCM. Shoot for a value of 4.6 volts at wide-open throttle. Be aware that the PCM will set an error code if it detects TPS voltage above 4.7 volts.

With the throttle stop ground, you can clearly see the difference in throttle blade angle. A word of caution: Take it easy with the grinding. Work slowly and deliberately, as it is easy to take away too much material. If that happens, you will have to weld additional material in place, or drill and tap the stop for a bolt to act as an adjustable stop. Either alternative is a pain, so do yourself a favor and be careful! Also, note the EGR provision directly behind the throttle plate. Clearly, this is detrimental to airflow and will be addressed shortly.

Coolant Bypass

The throttle-body coolant bypass mod is one that can be performed on virtually any multi-port fuel-injected GM engine. Engine coolant is routed through the throttle-body to keep it from icing up during inclement weather. Perhaps this was useful during development testing in the Arctic Circle, but it's no good for performance enthusiasts, as the hot throttle body casting warms the passing air and causes a decrease in engine efficiency. Check out the photos to see how easy it is to bypass the throttle body coolant circuit.

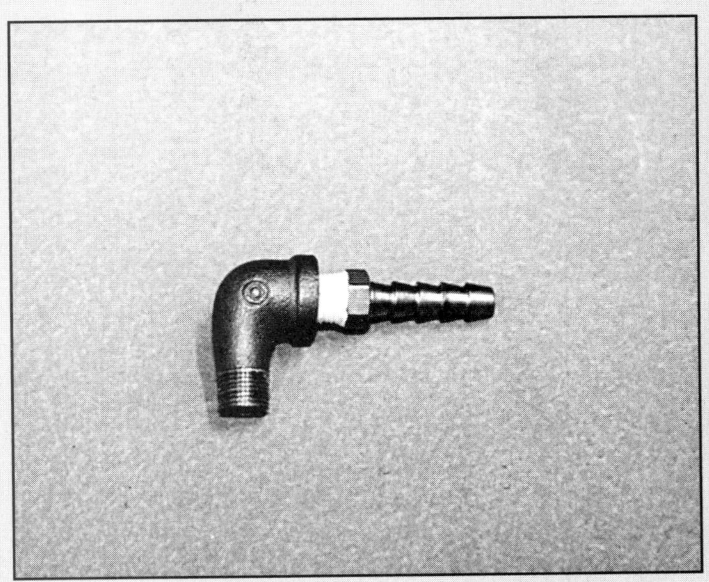

The throttle body cannot be simply bypassed with a single section of hose because the coolant line entering the throttle body is a smaller diameter than the line exiting it. Though several aftermarket vendors sell kits to perform this mod, you can definitely go the cheapskate route and build your own. Chances are that a quick rummage through your junk drawer will supply you with the fittings needed to make up an adapter like this. Wrap the threads with Teflon tape to prevent leaks and be sure to use hose clamps to secure the coolant lines to the fittings. Reuse of the stock clamps poses no problem.

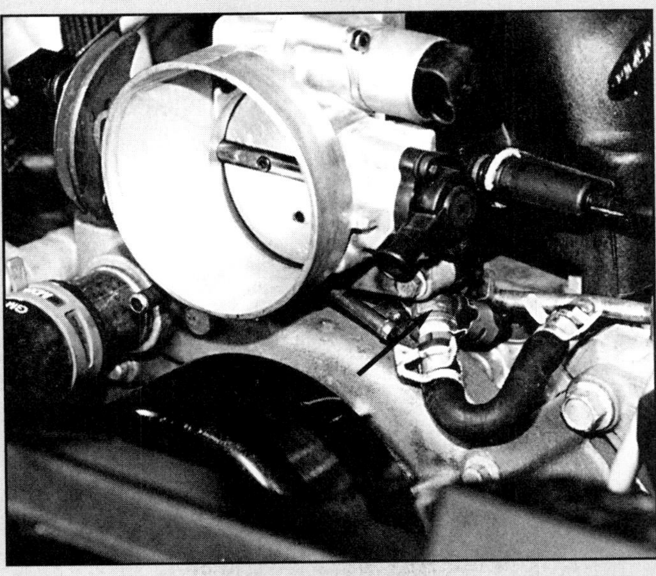

Here, you can see the adapter fitted between the two lines. The line from the driver's side of the throttle body can be seen in the foreground. The coolant line was rerouted beneath and behind the throttle body and clamped to the adapter fitting. Though you needn't drain the cooling system before performing this mod, you should definitely allow the engine to cool completely. You will also want to have a few rags to catch the coolant that will leak from the lines and throttle-body while you work.

To finish this mod, simply route the line behind the serpentine belt tensioner, being careful to keep it out of harm's way. If the line to the radiator on your car is not long enough to reach beneath the throttle body and behind the tensioner, a trip to the parts store for a longer section of tubing is in order. Keep this line away from the pulleys at all costs! As an extra precaution, secure the line in place with a tie-wrap or two.

3

Power Adders

In every group of automotive enthusiasts, there are always guys that can never get enough power, the ones for who overkill is barely adequate. They usually look to a power adder for their fix. The key to producing more power from the LS1 or any internal combustion engine is to burn additional oxygen and fuel during each combustion cycle. This can be accomplished in two ways. The first is by adding extra oxygen in a gaseous form as is done with nitrous oxide injection. The second is by compressing the incoming air charge as is done in turbocharging or supercharging. These methods are known as *power adders*.

Nitrous Oxide

Once the red-headed stepchild of power adders, nitrous oxide technology has come a long way in the last 10 years. Gone are the days of obscure black-and-white, postage stamp–sized classified ads buried at the back of *Popular Hot Rodding*. Nitrous is now big business. The bottom line is that there simply is no more cost-effective way of building horsepower than with N2O.

It wasn't too long ago that nitrous oxide injection suffered from a somewhat dubious reputation. Conventional wisdom held that nitrous was just about the fastest way to blow an engine. But these days, countless enthusiasts use nitrous regularly with nothing but lightning-quick time slips—the only thing breaking are e.t. records. Today's nitrous systems are so well designed and carefully calibrated, that engine failures have become comparatively rare. This is not to say that people don't still grenade engines when injecting nitrous, but the problem can almost always be traced to some fault with the installation or tuning. Nitrous is so effective at making power, many enthusiasts get a little greedy and can't resist the urge to up the hit, and this is where they get into trouble. By their reasoning, if some is good, lots must be great. So they jet it up a couple of notches and bang! The fun is over. That need not be the case, as a little common sense will go a long way toward ensuring long-term, trouble-free fun with the gas.

Nitrous oxide is a favorite among the LS1 power-adder crowd because it is relatively inexpensive and makes great power!

Nitrous oxide is stored in liquid form at high pressure, usually in a 10-pound bottle like this.

What is N2O?

Now that you know you want it, you may be asking: what exactly is nitrous? Nitrous oxide (N2O) is a colorless, oxidizing liquefied gas with a sweet, pleasant odor and taste. It is commonly used as a tranquilizer. This effect is preceded by mild hysteria, sometimes with laughter, thereby earning the name "laughing gas." Note that nitrous oxide destined for automotive use has a sulfur additive, which will produce some less enjoyable side effects, so don't sniff it! The product is stable and inert at room temperature. While classified by the U. S. Department of Transportation as a non-flammable gas, nitrous oxide will support combustion and may detonate at temperatures in excess of 1202° F.

Nitrous oxide is stored in a highly pressurized bottle, usually in the cargo area of the vehicle. The system is controlled by electrical solenoids, which when energized, allow nitrous (and in the case of a "wet" system, supplemental fuel) to flow to the point of injection. All systems use "jets" to control the amount of nitrous and fuel injected into the engine.

Nitrous oxide injection provides additional power in three ways. When nitrous oxide is heated to 572° F, it breaks down and releases its oxygen molecule into the charge mixture, allowing extra fuel to be burned. The second effect is by cooling the intake charge. Since the nitrous is stored in the bottle in liquid form (usually at about 1000 PSI), it expands into a gas when released into the manifold. For those of you who have a physics background, you'll know that this will cause the temperature to drop considerably, which will impart an increase in density. A typical nitrous system will drop the intake temperature by about 60–80 degrees. Actually, there is a third and often overlooked benefit to nitrous injection, and that is that the nitrogen molecules released during the breakdown process helps to buffer the increased cylinder pressure created. This improves the efficiency with which the oxygen/fuel mixture will burn.

Wet or Dry?

Many companies produce nitrous kits for the LS1, including Texas

"Wet kits" inject additional fuel with the nitrous oxide.

Texas Nitrous Technology's Power Ring wet injection kit uses two nozzles mounted in a machined aluminum collar between the MAF and throttle body.

Nitrous Technology, Nitrous Oxide Systems and Nitrous Express. Systems for injecting nitrous oxide into the LS1 are available in two distinct types: Wet systems, which inject additional fuel independent of the existing fuel delivery system; and dry systems that rely on the stock fuel system to supply additional fuel. Though each type of system has its own set of pros and cons, neither really has a clear-cut advantage.

Wet Systems—Wet nitrous systems typically inject both nitrous and fuel into the induction system via a plate or nozzle upstream of the throttle body. The biggest advantage to this type of system is that the nitrous/fuel proportions are consistent and can be tuned with separate nitrous and fuel jets.

One drawback to a wet system is that since the LS1 intake manifold wasn't intended to flow fuel, it does so rather inefficiently. This can result in fuel precipitating out of the mixture and forming puddles in the manifold. Second, the nitrous solenoid may need more frequent maintenance as the seals deteriorate due to prolonged exposure to fuel vapors. Lastly, if the nitrous pressure in the bottle rises too high, it may cause the air/fuel mixture to lean out. Conversely, as bottle pressure drops, the engine mixture will go rich.

Completely stock engines with stock fuel systems will comfortably support a 100-horsepower dose of nitrous delivered by a wet system. Installation of a high-performance in-tank fuel pump, such as a Walbro 340, increases the safe tolerance range to 150 horsepower. A built engine with forged internals can take 400 horsepower or more, but beyond that block reliability becomes an issue.

Dry Systems—The beauty of a dry nitrous injection system is its simplicity. Dry systems differ from their wet counterparts in that they inject nitrous oxide only, and rely on the PCM to supply supplemental fuel

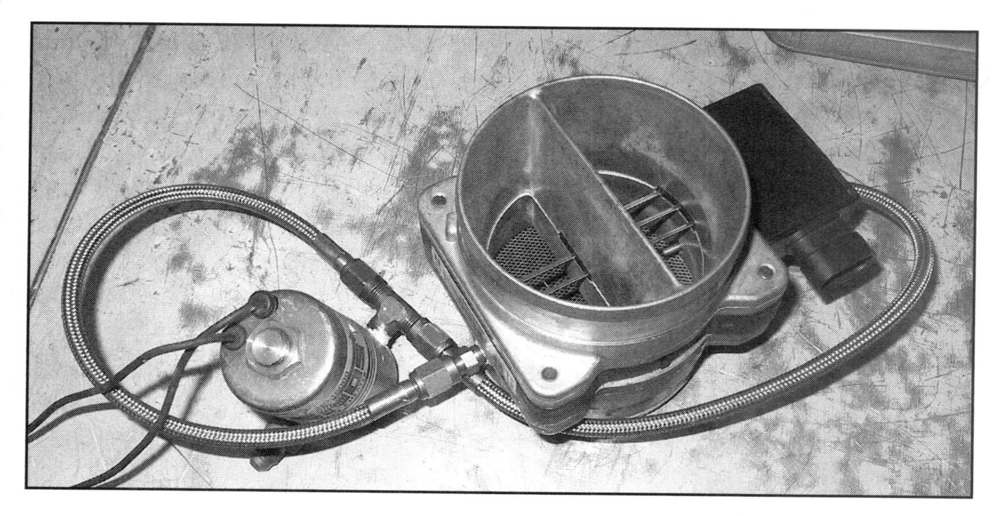

A "dry" kit injects nitrous oxide only, relying on the engine's PCM to supply additional fuel. This simplicity makes dry systems easy to install and very stealthy.

Direct port nitrous injection systems offer unparalleled tuneability because each cylinder has its own nitrous and fuel jet.

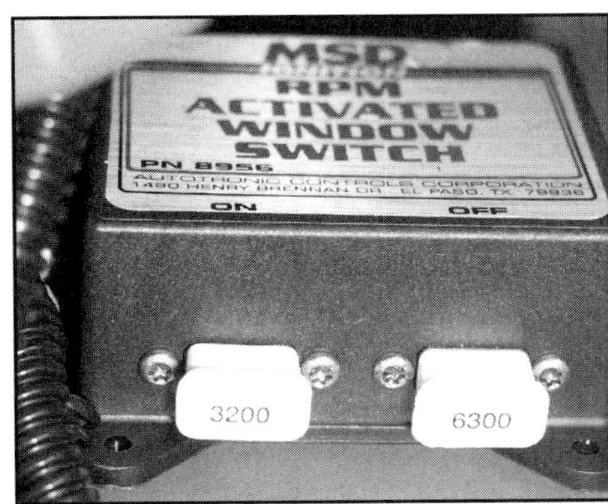

A window switch such as this one from MSD allows the nitrous system to engage only when the engine is in the user-defined rpm range, set by plug-in pills.

by increasing injector-pulse width. This is done by injecting the nitrous before the mass airflow sensor, which reads it as extra airflow. This increases the amount of fuel delivered by the injectors at the PCM's direction.

Benefits of a dry system include easier installation because it requires no plumbing for additional fuel delivery. Because fuel does not flow through the intake manifold, the chance of it puddling is eliminated. As the nitrous pressure in the bottle fluctuates, the amount of fuel delivered will change proportionately since the PCM is using information from the MAF to add the necessary fuel. A bonus feature for the street racer crowd is that the simplicity of a dry system makes it easy to hide.

Dry systems' main problem is that it is dependent on the fuel injectors for supplemental fuel delivery. If the injectors are not able to supply the fuel dictated by the PCM, the mixture will go lean. Therefore, injector sizing is critical for engines using a dry system. Another consideration is that the amount of nitrous delivered to each cylinder can vary slightly because it may not be evenly distributed in the intake manifold. However, the fuel will be the same at each cylinder, assuming the injectors are healthy. This can leave some cylinders lean while others run rich. If you plan to run more than a 100 horsepower hit, it is important to upgrade the fuel pump. If you're going over 150, you should consider a move to larger injectors. Completely stock engines with stock fuel systems will comfortably support a 100-horsepower dose of nitrous delivered by a dry system. Again, installation of a high-performance in-tank fuel pump increases the safe tolerance range to 150 horsepower, when stock injectors become a limiting factor. Installation of 36 pound per hour injectors can stretch the dry system to 200-plus horsepower.

Direct Port Systems

As you can imagine, it is conceivable that an engine running either style of nitrous system discussed previously could have some cylinders running lean while others are running rich. The best way to combat this problem is a direct-port, wet-style nitrous injection system. In this type of system, the fuel and nitrous jets are housed in a common nozzle, one per cylinder. Each fuel and nitrous jet are able to be tuned independent of each other and other cylinders. Direct-port setups offer the ultimate in tuneability, but with that tuneability comes complexity, which means a rather large price tag.

As with the previous nitrous injection systems, a stock LS1 can usually handle about a 150 horsepower dose of direct port injected nitrous without any internal engine modifications. Does that mean that there aren't people out there spraying more than double that amount? Well, no, but those people are playing with fire. The envelope can certainly be stretched beyond the numbers given here, but proper modifications to the fuel and ignition curves are a bit of insurance. In the end, it is up to you to decide which type of nitrous oxide injection system and how much to spray.

Window Switches

If you plan to run nitrous oxide injection, a window switch is the

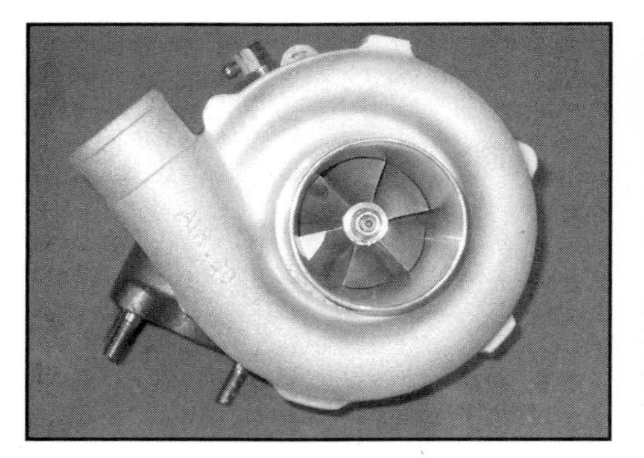

Turbochargers are a super-efficient way to increase the power of the LS1.

The turbine wheel is driven by exhaust gas and housed in a cast iron scroll as shown here.

A turbo's compressor housing is usually cast aluminum.

closest thing to an insurance policy you can buy for your engine. A window switch allows the nitrous system to function only when the engine rpm is within a user-defined range. If nitrous oxide is injected at too low an rpm, the tremendous cylinder pressure can be enough to bend connecting rods. Spray the motor while bouncing it off the rev limiter, and you will grenade it for sure. Don't think it will happen to you? Well, have you ever missed a shift or broken a drivetrain component? Get a window switch.

Turbochargers

A turbocharger makes the air/fuel mixture more combustible by compressing the air molecules, making the charge denser and stuffing it into the engine's chambers. It does this by pumping more air than the engine can use, resulting in boost. A turbo serves the same purpose as a supercharger, but it is more efficient, because there isn't the parasitic horsepower loss that results from a supercharger's drive belt.

Turbocharger Components

Turbochargers are composed of an exhaust gas–driven turbine section and a radial air-compressor section. These wheels are enclosed in cast housings and mounted at opposite ends of a common shaft. The shaft is enclosed and supported by a center bearing housing, to which the compressor and turbine housings are fastened. A typical turbocharger can spin at speeds in excess of 100,000 rpm.

Turbine—The turbine section is composed of a cast turbine wheel, a wheel heat shroud and a cast iron scroll housing. It is a radial-inflow device, meaning that exhaust gas flows inward, past the turbine wheel blades, and exits out the side of the

A wastegate can bypass exhaust gas from the turbine section of the turbo in order to limit boost.

Single turbo systems like this one from LS1 Motorsports can make over 900 horsepower on a low-compression engine built with forged internals.

Twin turbos can make just as much power with less boost lag, but are substantially more complicated. This kit is available from Strope Speed Shop in Washington, PA. 724-228-1166. (Courtesy Strope Speed Shop)

housing. Engine exhaust gas is directed through the exhaust manifold into the turbine housing. The exhaust gas pressure and heat energy extracted from the gas forces the turbine wheel to rotate, which drives the compressor wheel.

Compressor—The compressor section is composed of a cast compressor wheel, backing plate, and a cast aluminum housing. It is a radial-outflow device, meaning that the air flows outward, past the wheel blades, and exits at the outer diameter of the housing. The rotating compressor wheel draws ambient air through the engine's air filtration system. Its blades accelerate and expel the air into the compressor housing where it is compressed and directed through ducting to the intake manifold. If so equipped, an intercooler is placed between the compressor outlet and the throttle body.

Center Section—The compressor and turbine wheels' shaft is supported by a precision bearing system located in the center housing of the turbo. The bearings must position the wheels as closely as possible to the contour of the end housings to maximize efficiency. Key to the wheel positioning is the engine oil that fills and pressurizes the clearances between the center housing bore, bearings, and shaft. Seal systems separate the center housing from both the turbine and compressor stage. The seals restrict oil from entering the compressor and turbine areas and reduce the flow of gases from those areas into the center housing.

Boost Control—Wastegates control boost by allowing exhaust pressure to bypass the turbine housing, reducing turbine speed. They can be an integral part of the turbine housing or mounted remotely, such as on a header collector. When opened, excess exhaust pressure is released from the turbine housing, directed to the exhaust system and expelled into the atmosphere.

The goals set for power will determine the amount of boost required to deliver the needed volume

A centrifugal head unit such as this ATI ProCharger D1 can supply enough air (1400 cfm) to support over 900 horsepower.

This bolt-on ATI ProCharger system for LS1-powered Corvettes features a self-contained P-1SC-1 head unit and twin intercoolers. It is capable of producing well over 500 horsepower on stock engines running pump gas.

of air to the engine. Combinations of wheel size, wheel speed and housing size accomplish this. A stock LS1 engine will generally tolerate up to seven pounds of boost so long as there is an intercooler in use (more on that later). Be warned that running higher levels of boost is possible, but not recommended. If the mixture leans out and detonates, chances of burning a piston are good. If you are screwing together a purpose-built engine, the sky is the limit when it comes to boost, as long as the proper precautions are taken. Some of the items required to run big boost include lowered compression ratio, forged pistons, upgraded head bolts or studs.

Single or Twins?

A turbo system can be configured to use single or multiple units (usually twins), and neither type has a clear-cut advantage. Single turbo kits like those available from LS1 Motorsports and Turbo Technologies, are simpler, a little easier to install, and a little less costly. In a single turbo setup, a larger turbo is required to provide the same airflow as a twin turbo system. The larger the turbo, the more boost lag is likely to occur. The twin turbos main advantage is quicker spool time because the turbos can be much smaller yet combined, flow the same amount of air as a large single. The downside is increased complexity and more components to buy.

Superchargers

Like a turbo, a supercharger (also commonly known as a blower) is a positive displacement air pump that pressurizes the intake manifold. Unlike a turbo, a supercharger is driven directly from the crankshaft by a belt. This positive connection yields instant response, in contrast to turbochargers, which must spin up to speed as the flow of exhaust gas increases. The supercharger is matched to an engine by its displacement and belt ratio, and can provide boost at any engine speed. If properly matched, this can make for an extremely responsive package with no lag in boost production. There are currently two types of superchargers available for the LS1, centrifugal and screw-type.

Centrifugal Superchargers

Centrifugal superchargers look much like a turbo, compressing air inside the head unit using an impeller before discharging it out of a scroll to the engine. Centrifugal supercharging is definitely one of the more user-friendly ways to supercharge an engine. The ability to change the impeller sizes and to spin the impeller at different speeds allows flexibility in the engine's power curve. Belt slippage can be a problem at high power levels, but can be combated through the use of a cogged belt setup. While cogged pulleys and belts solve slippage problems, they are impractical for street use. Several manufacturers offer centrifugal supercharger systems for the LS1, including ATI Procharger, Powerdyne and Vortech.

Screw-Type Superchargers

Screw-type superchargers use an axial-flow design that compresses the

Magnacharger's screw type supercharger makes well over 450 horsepower as installed on a stock Corvette.

Intercoolers improve the efficiency of turbos and blowers by removing heat produced during compression.

air as it moves between twin screws, allowing it to create positive pressure without excessively heating the air. They produce power right from idle and offer a high degree of reliability, making them an excellent choice for trucks used to tow. As for their installation on cars, cowl clearance issues on the F-bodies currently limit screw-type supercharger use to Corvettes. Even then, a modified hood is required. Expect a screw -type supercharger to make between 425 and 450 rear wheel horsepower on an otherwise stock car.

Intercoolers

An intercooler is a simple heat exchanger designed to remove heat from the compressed air coming from a supercharger or turbo. The reduction in air temperature increases the density of the air, which increases the engine's ability to make more horsepower and torque. The most significant advantage of intercooling is that it increases the detonation threshold by delivering a cooler, denser air charge to the engine's combustion chamber. This means the engine will tolerate more boost, more ignition advance, or lower octane fuel before experiencing detonation.

Intercoolers do create an airflow restriction that means a slight reduction in boost at the manifold, and can also cause the engine to run leaner due to the denser air charge. These small setbacks are far outweighed by the benefits mentioned previously. Boost pressure can be increased (in fact, you'll probably want to run substantially more boost than you would with a non-intercooled system) and the air/fuel ratio can be tweaked through the use of larger fuel injectors, a larger fuel pump, a MAF Translator or changes to the PCM's calibration.

Aftercoolers

Aftercoolers, or air-to-water intercoolers, are cooled by water instead of air. The benefit of an aftercooler is that the charge air can be cooled more than in a traditional air/air intercooler if very cold water and ice are used. Some aftercoolers are able to chill the air below ambient air temperatures even after the supercharger has compressed it. The drawback is that with time, the water

Vortech's aftercooler assembly includes pump, wiring, surge tank, reservoir and a front mount heat exchanger for effective street operation

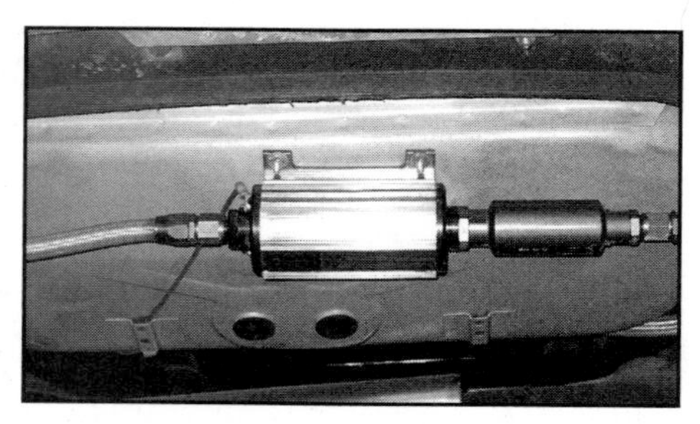

Fuel system upgrades are a vital component of any power adder installation.

will heat up to the temperature of the air passing through it, and its ability to cool incoming air goes away. For drag racing applications, aftercoolers work very well because they only need to work for short periods of time. For milder racing and street applications air-to-air intercoolers are far more practical as their ability to cool incoming air is not reduced with time.

Obviously, intercoolers of either type work only in conjunction with forced induction engines where there is a substantial difference in temperature between the air entering the engine and the cooling medium. Because turbos and superchargers heat the air as they compress it, it is possible for there to be a very large temperature difference between the intercooler (ambient air temperature—80F degrees or so) and the comp-ressed air (200F–350F degrees). Higher boost will create a higher discharge air temperature, so as you increase your boost, the effects of the intercooler become more and more beneficial.

Fuel Delivery and Power Adders

One of the biggest reasons that an engine goes south when using a power adder is that the air/fuel mixture goes lean. The fuel delivery system for any power adder engine needs to be up to the task of delivering the extra fuel to the engine.

The stock in-tank fuel pump is usually sufficient for nitrous systems limited to 100 horsepower. Anything over that amount, and I'd recommend upgrading the fuel pump or adding an extra in-line pump to supplement the in-tank one, depending on the level of additional horsepower. You may not need it, but it's cheap insurance compared to the price of a complete engine rebuild. A clean fuel filter is an important maintenance item; it's too cheap not to replace at regular intervals. As the power level increases, more upgrades to the fuel system will be needed. These modifications are addressed in chapter 9.

Ignition and Power Adders

Power adder engines require a few changes from the stock setup to perform well, specifically the plug type and gap, and the ignition timing. The stock platinum sparkplugs in the LS1s are great for daily driving, but lousy when a power adder is in use. Because they retain a lot of heat, they will cause detonation when subjected to the increased cylinder pressure induced by nitrous or boost. Therefore, a conventional plug that is one or two heat ranges colder than stock is a good idea. You should also close the plug gap down to about .035-inches, so that when the mixture ignites it does not blow out the spark. Reducing the timing advance reduces the likelihood of detonation. A good rule of thumb is to reduce advance about 1 to 1.5 degrees for every 50 HP of nitrous or 4 pounds of boost. Ignition timing is controlled by the PCM, and you will find more information on this topic in chapter 10.

Upgraded rod bolts, such as these from ARP, are an excellent upgrade for almost any engine.

In order to check rod-bearing clearance, the bearings were put in place and the caps were installed. Once clamped in the rod vise with the bolts torqued to 75 lb-ft., the inner bearing diameter could be checked. In this case, they measured in at 2.1023-inches. Repeat this for the seven remaining rods. The rod journal diameter on the Lunati crankshaft was a perfect 2.0995-inches on all journals. The difference in diameters is known as clearance, and was right on the money at .0028-inches.

Recommendations

Stock LS1 connecting rods are quite suitable for most builds that retains stock length rods. It is good practice to Magnaflux inspect them before resizing as necessary. Some may choose to polish the beams to reduce the chance of stress risers. This is

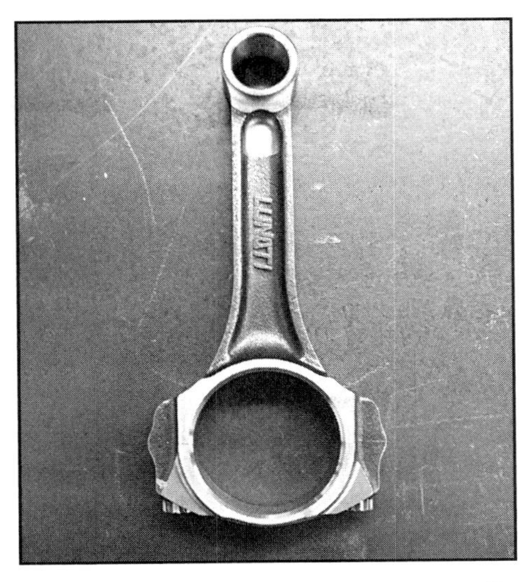

Aftermarket connecting rods are a necessity if increasing the stroke of your LS1.

Ideal rod bearing clearance is .0028- to .0030-inches. If the bearing clearances are too loose you can either mill the caps or use oversized (thicker) bearings. If they are too tight you can use undersized bearings to gain additional clearance. Remember, bearing clearance impacts oil supply, and if your rod bearings are not fed properly, they will not last.

unnecessary as the rods are shot-peened by the factory for this reason.

Rod Bolts

If the OEM rods have an Achilles heel, it is the rod bolts. For simple assembly and mass reduction, the LS1 rods use 9mm cap screw rod bolts. Rod bolts for the 1997 to 2000 model years are particularly suspect. Though GM improved the design of the bolts in 2001, any engine beyond a stock rebuild should be treated to top quality fasteners such as those available from ARP. They are relatively inexpensive and are good insurance against a major engine catastrophe.

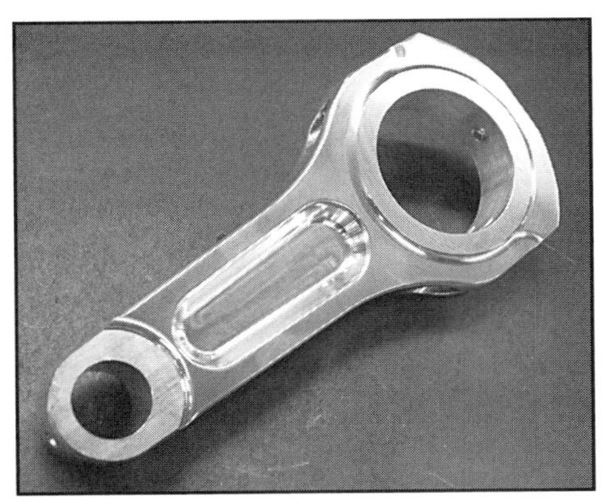

Aluminum rods are very sexy and very lightweight, but have a somewhat limited lifespan.

Forged and Billet Rods

Stroker cranks require shorter connecting rods to keep the pistons in their bores at the top of the stroke. Aftermarket steel rods are available in two types: billet and forged. Billet rods have a slight edge in both weight and strength, but are overkill for most engines. Forged rods are strong enough for all but the most extreme nitrous or forced induction application, and more affordable to boot. Lunati and Oliver are two companies offering top quality connecting rods for the LS1.

Aluminum Rods

Aluminum rods significantly reduce reciprocating weight by as much as 150 grams each, allowing the engine to rev quicker. They also have a tendency to absorb shock, so the use of these rods can lessen the beating transmitted to the bearings should the detonation occur. However, aluminum is prone to stretch and therefore rods of this type need periodic replace-

Stock pistons are okay for stock applications, but their eutectic construction makes them fairly susceptible to damage from detonation.

LS1 pistons have a very short compression height of 1.339-inches. Low-tension rings are installed to reduce friction.

JE, Lunati and Ross are a few companies that manufacture top-quality forged aluminum slugs for the LS1. Most are available with valve reliefs that provide the piston-to-valve clearance necessary when running long strokes, aggressive compression ratios and cam profiles.

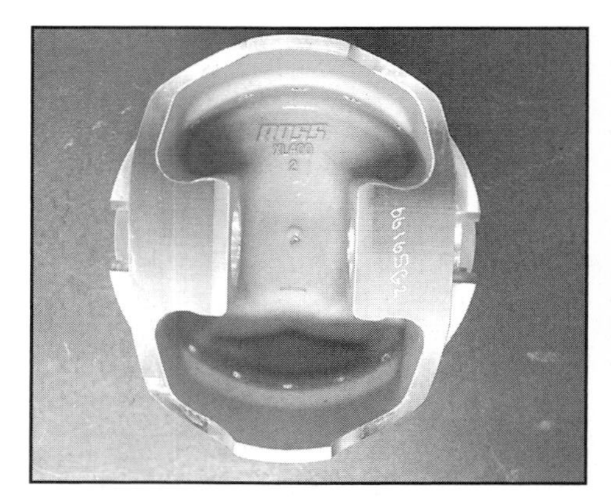

The underside of a typical Ross forged piston.

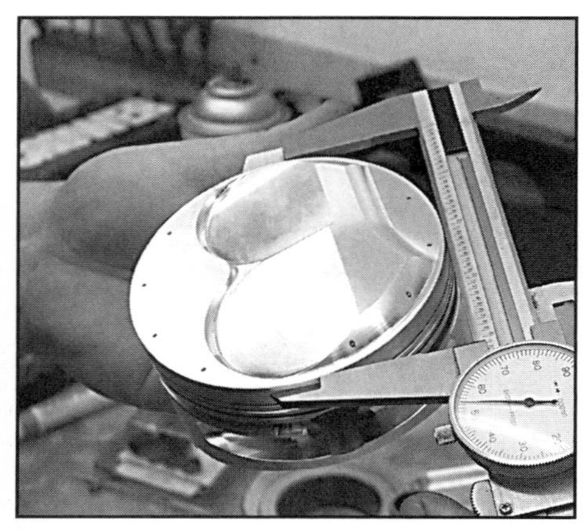

It is unusual to find improperly packaged pistons, but it is well worth the time to verify their diameter. These Ross pistons measured 4.076-inches, which produces a clearance of .0045-inches in the previously measured 4.0805-inch bores. Final compression ratio, when combined with the 64cc chamber cylinder heads, worked out to a pump gas friendly 10.9:1.

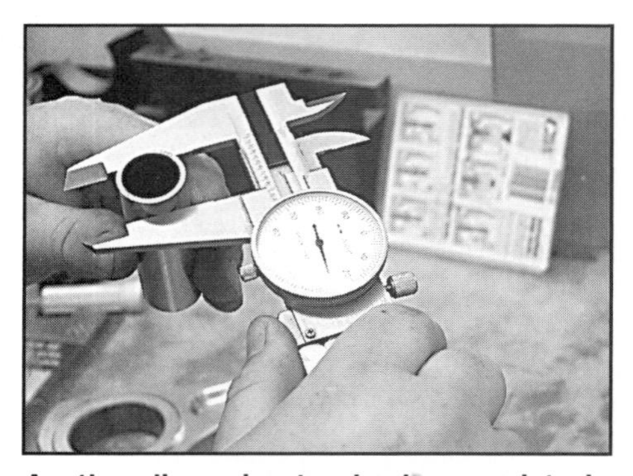

Another dimension to check are wrist pin diameter and the wrist pin bores in both the rods and pistons. These wrist pins are .9270-inch in diameter, so the rods were machined to .9276-inches, which provided .0006-inches of clearance.

ment. So although they cost about the same as billet steel rods, the constant replacement required makes them practical only for a maximum-effort race engine.

Pistons

The LS1's eutectic aluminum pistons have flat tops with no valve reliefs and feature a closed skirt, strutless design. Because process controls at the block machining stage are tighter, bore variation was significantly reduced. This means piston expansion control was less of an issue, and the steel strut was eliminated making for a piston that is lighter and less expensive to manufacture. They have a compression height of 1.339-inches (34mm) and weigh 434 grams. The top land is .177-inches (4.5mm) thick to reduce crevice volume and hydrocarbon formation.

Forged Pistons

Pistons will obviously need to be changed if there is a bore diameter or crankshaft stroke change. There are also a handful of companies like Ross that offer custom pistons tailored to your specific needs. Several piston companies offer off-the-shelf forged replacement pistons. Forged pistons offer much improved resistance to detonation, but are more likely to exhibit piston slap upon start-up.

Piston Rings

The LS1's stock compression rings are .059-inches (1.5mm) thick. The first ring is barrel-faced molybdenum-filled 9254 steel. The second ring is cast iron, has a tapered face, and reversed twist. The oil control ring is conventional. All rings are low tension for reduced friction.

Lubrication System

The Gen-III block posed an interesting design problem with regards to the lubrication system. Its

Piston/Rod Assembly & Ring Gap

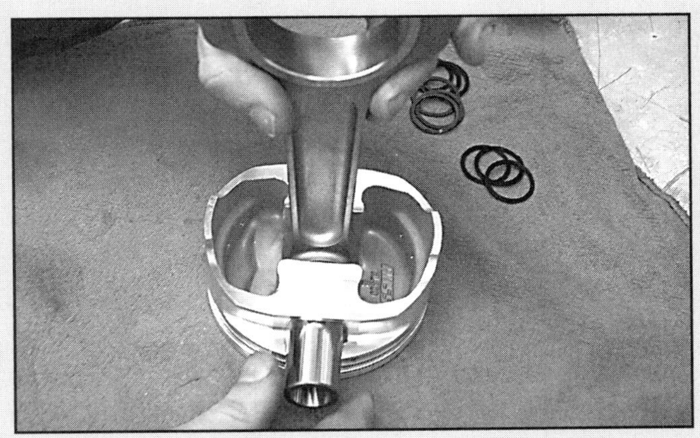

With the small end of the rods properly sized, it was time to assemble the rods and pistons. After a liberal dousing in clean engine oil, insert the wrist pin to hang the rods on the pistons.

Forged pistons are usually equipped with floating wrist pins. The wrist pins are retained by spiral locks, which snap into a groove machined in the piston.

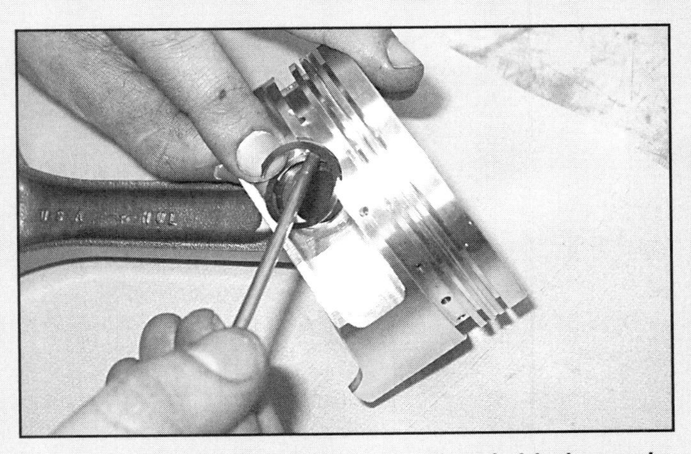

With the wrist pins in place, the double spiral locks can be installed. These locks are quite secure and a real bear to get back out should the need for disassembly arise.

These pistons are in order, ready for ring installation. But before installing them, the rings need to be properly gapped.

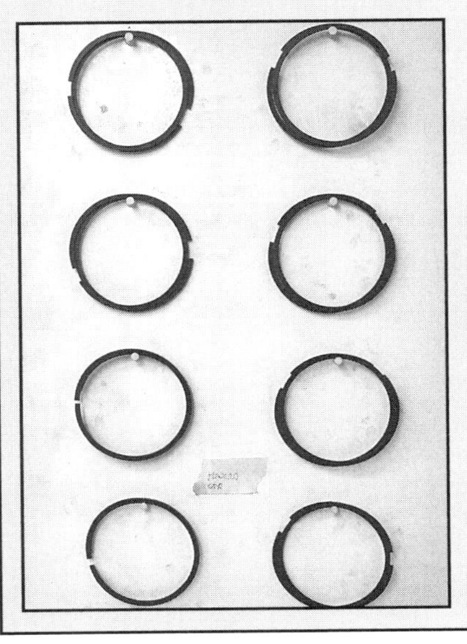

It is imperative that the rings be kept organized according to cylinder when their gaps are set. A peg board is one idea.

The key to proper ring gap is ensuring that the rings are seated squarely in the bore for gap measurement. Push the ring down with either a piston or ring compressor. The rings for this engine were Childs & Albert .043-inch Dura-moly top rings, 1/16-inch ductile iron second rings and a standard 3/16-inch oil ring.

Next, measure the ring end gap. Do this after the crank is installed and caps torqued, as there will be a small amount of distortion created by the main bearing fasteners, and that will affect this gap.

In order to adjust the ring gap, use a ring grinder (or file by hand if you are truly a masochist). Grind and remeasure until it reaches the specified gap. For this engine, the top ring and second ring were set up for .020- and .016-inch gaps, respectively. Do this for all of the rings, each in their own cylinder. Remember to keep the rings organized and labeled after setting the gaps.

With the rings gapped, each for a specific cylinder, they could be installed on the pistons either by hand as shown here or with a ring-expander tool.

Shortblock Final Assembly

Once you've checked dimensions, verified clearances, and prepared the subassemblies, it is time for the final assembly. Be sure to keep mated parts organized and have all fastener torque specifications on hand.

Some engine builders prefer the use of heavy engine oil to assembly lube. Assembly lube has been known to clog oil pickup screens at startup and destroy a new engine due to the lack of oil pressure. If you decide to use oil, don't be stingy with it during assembly. Be sure to apply it liberally to the pistons, rings and rod bearings prior to assembly. A copious amount should also be spread around the cylinder walls.

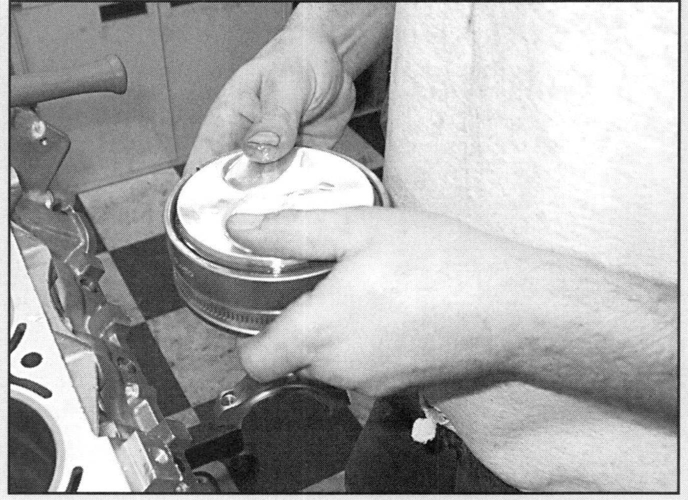

A ring compressor will be needed to insert the piston/ring/rod assembly into the cylinder bores. As the piston and rings move into the compressor, the built-in taper gradually compresses the ring to allow it to slide easily into the bore. It is critical to ensure the rings are clocked correctly, with the gaps 180° apart. Power robbing blow-by can otherwise be a problem.

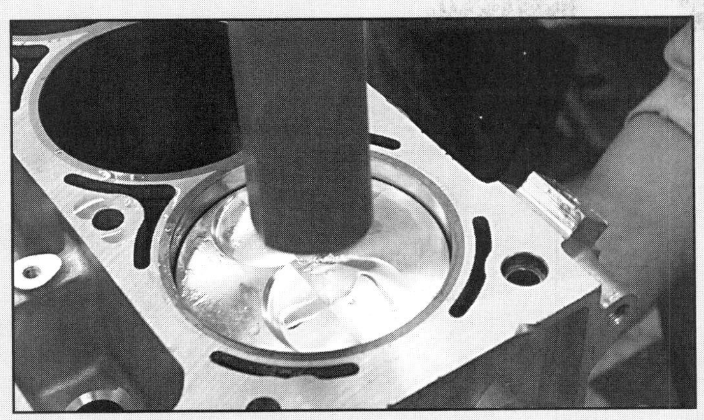

With the piston and rings started in the bore, remove the compressor and tap the piston the rest of the way down the bore. You can use the rubber-coated handle of a mallet, or if you want to get fancy, you can get a special dead blow hammer designed for the task, as shown here. As always, care should be taken to prevent the rod coming in contact with the cylinder wall.

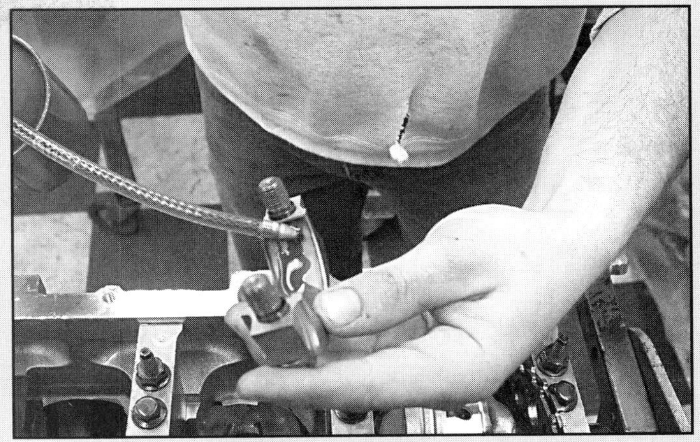

More oil should be applied to the rod cap bearing and rod bolts before assembly. The rod bolts were installed and torqued to 75 ft.-lbs. Be sure to spin the engine after each cap is torqued to check for binding.

If you are building a long-stroke LS1 like this one, the number eight piston requires a small amount of extra preparation. The bottom of the piston skirt must be clearanced to provide space for the reluctor wheel on the crankshaft.

After each pair of rods and pistons are installed and torqued, use a feeler gauge to measure the rod side clearance. This clearance depends upon if you are using steel or aluminum rods. Since this engine is utilizing steel rods, clearance of .015-inches is correct. Clearance with aluminum rods should be about .020-inches, but as always, verify these numbers with your rod supplier. If the clearance is too tight you can machine the rod.

Though not at their closest point, this shot shows how close the rod bolts pass the bottom of the cylinder bore, illustrating why block clearancing was necessary. Minimum allowable clearance between the rotating assembly and the block is .030-inches.

Piston deck height is the difference between the top of the piston at TDC and the block deck. As designed by GM, the LS1 has a negative deck-height figure, which means that the piston protrudes into the space surrounded by the head gasket. One of Powertrain's goals in combustion control was to decrease crevice volume, which is the squish volume between the flat, non-chambered, part of the head exposed to the bore, plus the volume between the piston and bore above the top ring. In the case of this engine, it is .0070-inches higher than the block deck, increasing compression. Such things as combustion-chamber volume, head-gasket thickness, and crevice volume determine final compression ratio.

There are currently no aftermarket oil pumps available for the LS1, so the stock unit should be ported to improve flow.

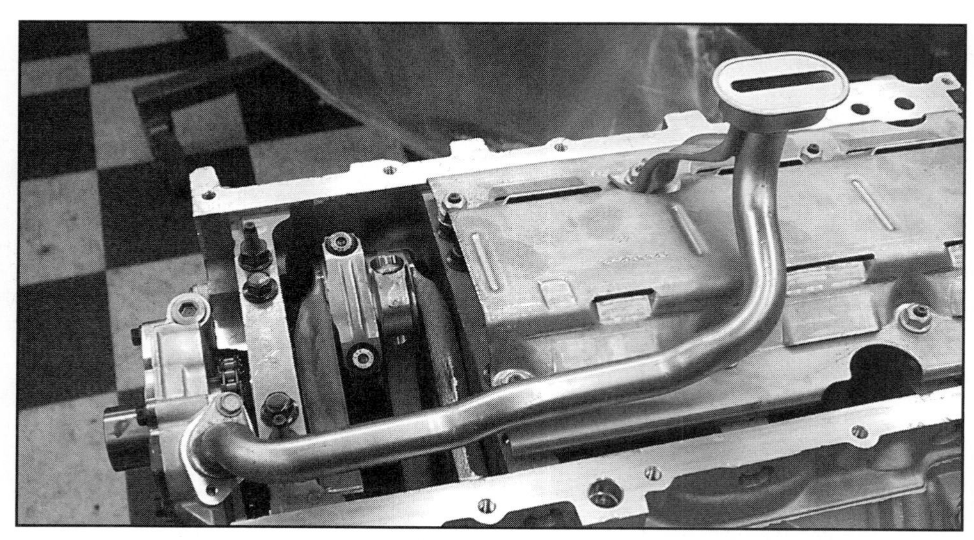

The shape and location of the pickup was developed to maintain oil flow under all vehicle operating conditions.

deep-skirted block, six-bolt main bearing caps and higher oil level (that results from the shallower oil pan) effectively divide the crankcase into four distinct bays. The problem with this is restricted transfer of air between bays as the pistons moved in their bores. At high rpm, the violent turbulence caused by this absence of pressure relief caused aeration of the oil and restricted oil drain-back from the upper end of the engine. The combination of oil foaming and poor drain-back degraded the oil supply to the point that at high degrees of lateral acceleration, oil in the pan could reach a 45° angle, uncovering the pickup. Sustained occurrence of these issues would result in loss of oil pressure when the oil pickup was sucking air. This problem was combated by adding 1.12-inch (28.5mm) ventilation holes to all the cylinder block bulkheads, .965-inch (24.5mm) diameter holes through #2, 3 and 4 crankshaft main journals, and side pods to the cylinder block, allowing air to flow around the main bearing caps.

Significant features of the LS1 oiling system include the front pump, rear filter arrangement and the LS1's main oil galley feeding the main bearings and the camshaft simultaneously. High rpm operation creates high bay-to-bay flow, crankshaft windage and the potential for oil aeration. GM addressed this by making internal ventilation changes, adding a crankshaft deflector and an integrated baffle in the oil pan. The new firing order increases the minimum oil film thickness by 13%. Low flow main bearings further reduce oil flow requirements (22.7 liters per minute at 6000 rpm) and increase overall system lubrication and pressure.

Oil Pump

The LS1 oil pump could not be driven from the camshaft due to constraints on overall engine length. For this reason, it is located behind the harmonic balancer on the nose of the crankshaft. It is isolated from the front cover to minimize oil pump noise transmission. The pump's compact and efficient Gerotor design is less complex and cheaper to manufacture. In spite of Powertrain's best efforts, it is no secret that the LS1has an oil supply problem that leads directly to starved bearings. A ported oil pump goes a long way toward minimizing oil starvation problems.

Several vendors such as Agostino Racing and MTI offer a ported oil pump for the LS1 engine that improves oil flow. Ported oil pumps increase oil-system efficiency by allowing smoother flow through the pump body. The pump is easily disassembled, so you should consider modifying it yourself. See the nearby sidebar for instructions.

Pump Pick-Up

The oil pump utilizes a .879-inch (22.35mm) pick-up tube to deliver oil to the pump. The pick-up tube screen is oval with its long axis parallel to the crankshaft centerline. The shape and location of the pickup was developed to maintain oil flow under all vehicle operating conditions.

Oil Pan

The LS1 oil pan is a lost foam casting made of 356-T6 aluminum and an integral structural component

How to Port an LS1 Oil Pump

As with porting cylinder heads, the idea behind porting an oil pump is to remove any ridges or sharp edges that are disruptive to flow. Here is a look inside a disassembled, but unmodified, stock pump.

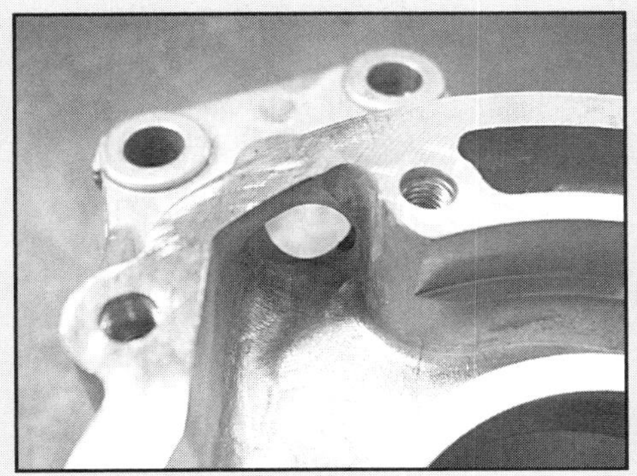

A few minutes with a grinder produces a much smoother oil flow path.

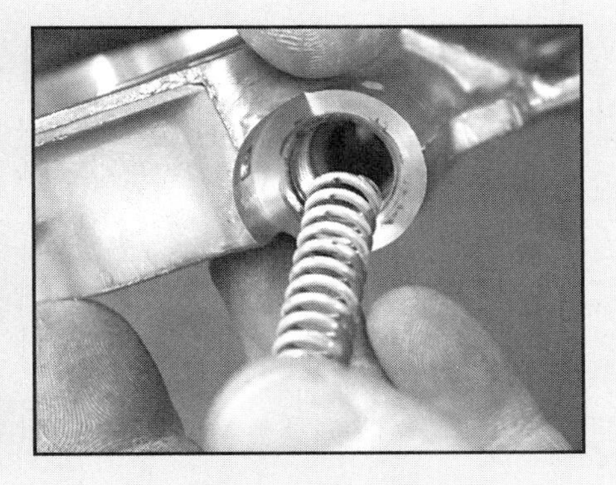

Thoroughly clean the pump housing before reassembly. Install the bypass spring in its bore.

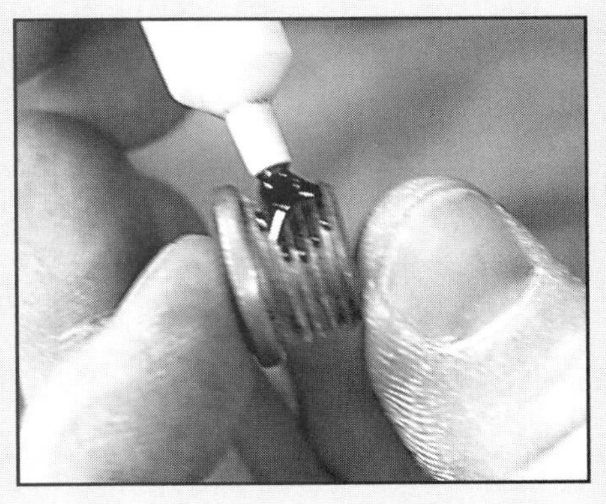

It is a good idea to apply a bit of Loc-Tite to the spring's pipe plug before threading it in place.

Tighten the plug carefully, torquing it to 12 N-m. Too much pressure will split the pump housing.

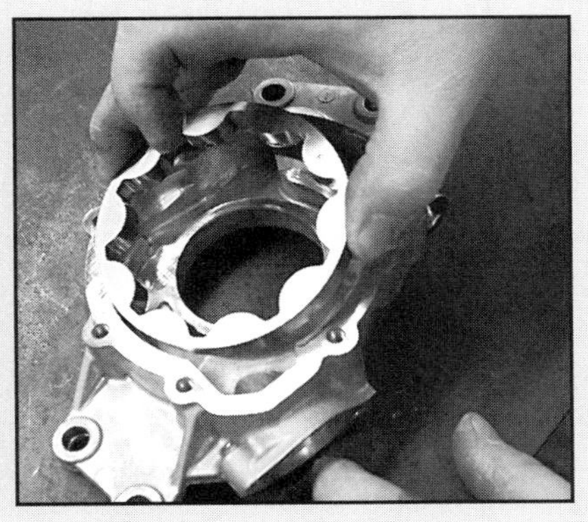

Next, insert the outer gear.

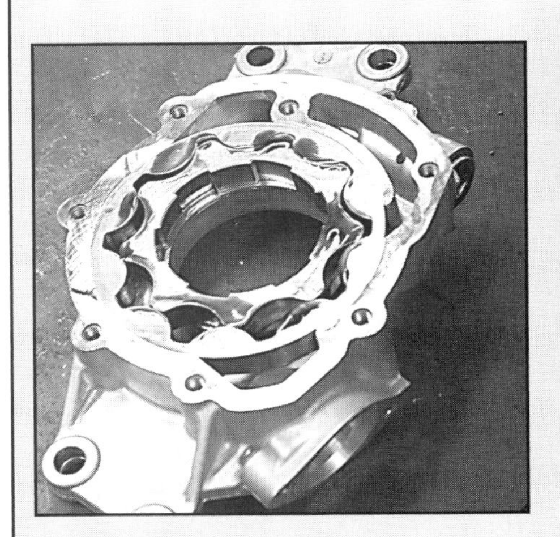

Coat the inner gear with a uniform coat of grease to protect it during initial start-up.

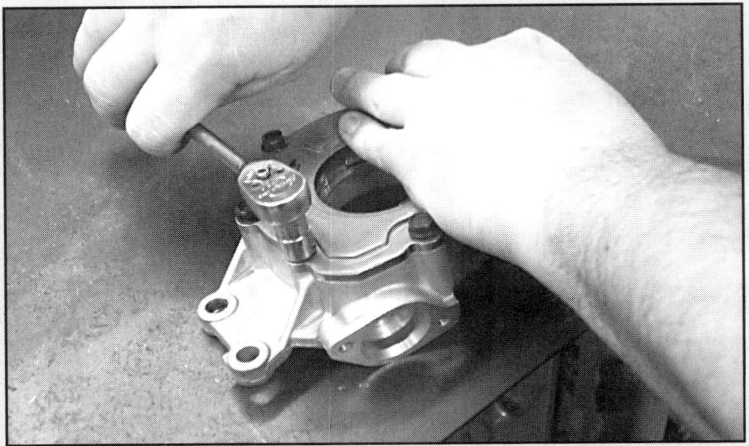

Replace the cover and tighten the fasteners accurately, torquing to 12 N-m. Again, do not overtighten.

The oil pickup tube mates to the oil pump body and is sealed with this neoprene O-ring.

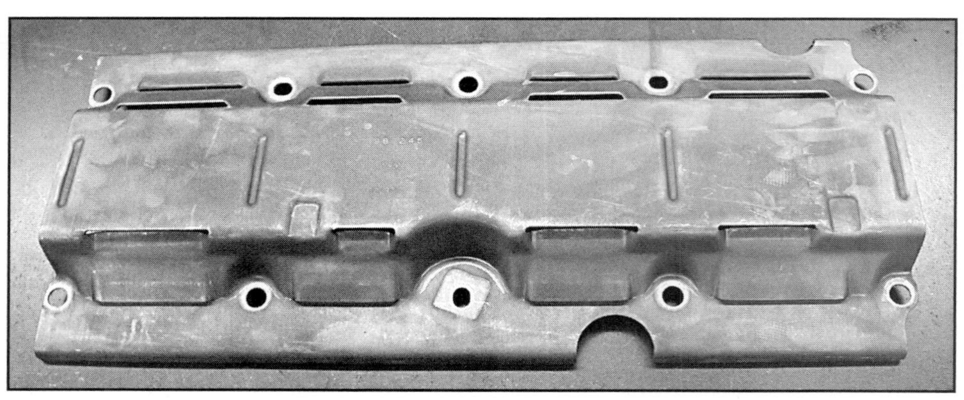

A windage tray is another tool designed to improve oil return to the sump.

The C5 oil pan has a unique extended sump that provides adequate oil capacity while allowing proper ground clearance for the low-slung 'Vette.

The windage tray is installed over the crankshaft main bearing caps.

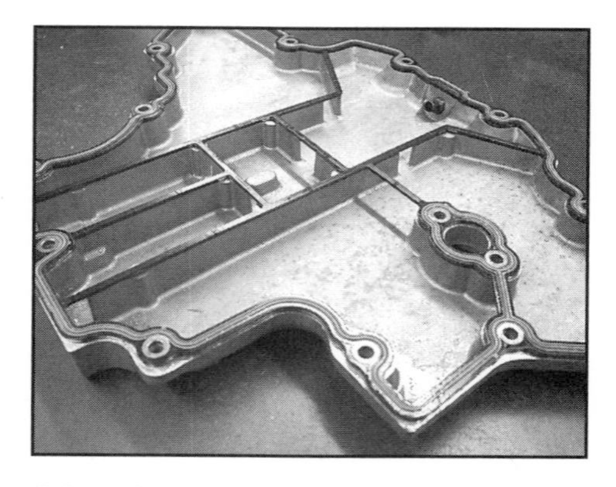

A look inside of a disassembled C5 pan reveals the compartmentalized sump with limited-flow windows intended to keep oil at the pickup during extreme maneuvers.

of the engine. Incorporated into the design is the oil filter mounting boss, drain plug opening, oil level sensor mounting bore, and oil pan baffle. The oil pan cover, oil temperature sender (for Corvette) and oil level sensor mount to the side of the oil pan. A shallow oil pan sump was required due to vehicle packaging constraints, but made it difficult to keep oil around the pick-up tube under extreme dynamic conditions.

To provide a constant supply of oil to the oil pick-up, reservoirs were added to each side of the pan. This allowed oil volume to be added to the pan while keeping the oil level below the crankshaft, preventing additional aeration of the oil. Dams were added to guide the oil to the pick-up with the help of strategically placed windows in the walls. A shelf at the front edge of the sump was added to keep oil in the sump during hard braking.

Baffles and Windage Tray

The stamped oil pan baffle is an integral part of the oil pan assembly. The baffle is located on a ledge in the pan, sealing the perimeter of the oil pan sump. The baffle maintains an area around the pick-up tube to prevent oil system starvation and aeration. The baffle diverts oil into the sump using several drain-back slots and a trough to direct oil into the cavity where the pick-up screen is located. The front edge of the baffle is turned upward to also assist in directing oil down into the sump upon

The alignment of the structural oil pan to the rear of the engine block and transmission bell housing is critical to achieving a perfect seal.

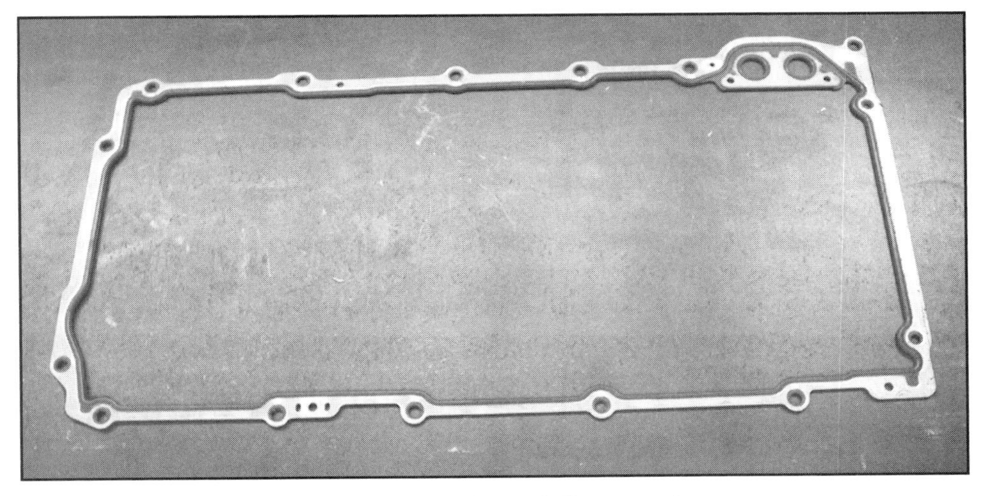

Gen-III oil pan gaskets feature aluminum carriers and silicone seals.

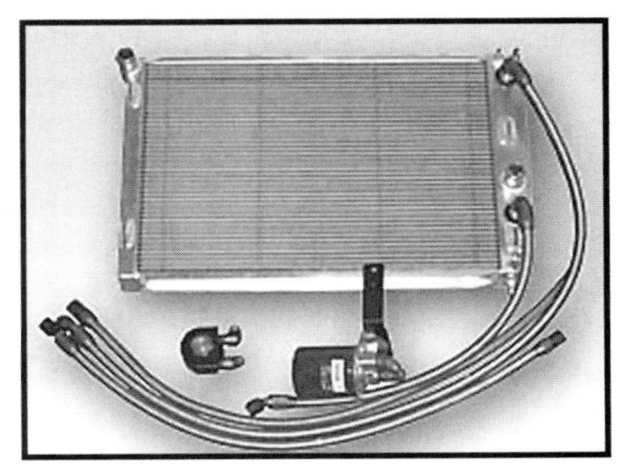

Doug Rippie Motorsports offers this high-end oil cooler system and aluminum radiator for road-racing C5s.

acceleration after hard braking.

The windage tray (deflector) mounts to the main bearing caps. Its job is to control windage, scrape oil from the crankshaft, facilitate drain-back and reduce aeration. Key performance factors of the deflector include size and location of the slots, gap between deflector and cylinder block, and proximity of the deflector to the crankshaft

Oil Cooling Systems

GM has deemed oil cooling systems unnecessary for the LS1, even the ultra-performance LS6. However, if extended high speed driving is part of your plan, an oil cooler may be a wise investment.

Gaskets and Seals

Powertrain designed many changes into the block and gaskets to enhance their ability to seal the engine. The deep skirt allows the oil pan to mount upon one surface instead of the twin fore and aft radii of past engines. A flat valley cover and carrier gasket with direct attachment seals the block valley enclosure points all around.

Controlled compression aluminum carrier gaskets are also used for the front and rear cover and the oil pan. Carrier gaskets are superior because the sealing point, which is silicon, is not subject to direct compression force. The wide aluminum surface for the gasket carries the majority of the load. Although minor variations in the torquing process may occur, uniform sealing is still achievable because of the dense, compression limiting aluminum. This also prevents any significant creep relaxation, which would result in a loss of fastener load, and ultimately, a leak.

A polytetrafluoroethylene (PTFE, Teflon) lip on the rear main crankshaft seal ensures long life. The second inboard lip eliminates false static leak test failures during assembly and is made of Viton. The purpose of the excluder shoulder is to provide a labyrinth seal to divert larger debris. Like the rear seal, the front main seals key component is a PTFE lip. This sealing lip increases seat-to-shaft sealing area with a wide lay-down design. When rebuilding your LS1, installation of all new GM gaskets and seals virtually guarantees a clean garage floor.

The rear main seal is press-fit in the rear engine cover, as is the seal in the timing cover at the front of the engine. Replacing these seals prior to reinstallation of the covers is cheap insurance against an annoying oil leak.

The LS1 water pump is cast aluminum and features an integral thermostat housing.

The Cooling System

The engine cooling circuit of the LS1 is arranged in a conventional flow path. Coolant is moved from the pump to the cylinder case, traversing the head gaskets before flowing through the cylinder heads and exiting the engine. This bottom-up flow maximizes the vapor purging capability of the circuit.

The LS1 employs an aluminum water pump, centrally located at the front of the engine, designed to provide equal distribution of coolant to both banks of the cylinder case. The coolant pump is belt driven and the impeller is doubly shrouded, improving pumping efficiency and total pump output. The pump housing incorporates a coolant crossover and bypass circuit.

Thermostat

The thermostat's inlet location eliminates thermal cycling common with outlet side designs. A large, spring-loaded bypass valve rests on the bypass circuit seat when the thermostat main valve is closed, maximizing coolant flow available to the heater core at low engine speeds. The bypass control is calibrated to provide nearly constant pump flow, regardless of thermostat valve position, providing more uniform heat transfer characteristics and temperature control. The coolant returning from the heater circuit is directed at the thermostat wax motor which biases the controlled coolant temperature slightly warmer when the heater core is extracting energy, thereby improving heater perfor-mance.

Modification

A wax pellet in the thermostat body controls the LS1 thermostat. As it heats up, the pellet (known technically as a wax motor) expands and acts directly on the thermostat piston, forcing it open and allowing coolant to pass. By placing a spacer in the housing between the piston and motor, the thermostat will open sooner, thereby keeping coolant temperature lower. One advantage to this is a cooler running engine that is less likely to detonate. For those of you who would rather deal with the sticky, stinky coolant just once, you may want to perform this mod at the same time as the throttle body coolant bypass modification found in chapter 2.

A separate vent circuit is attached to the front and rear inboard corners of

While there are a few cooler, aftermarket thermostats available for the LS1, almost anyone can modify the stocker to the same effect.

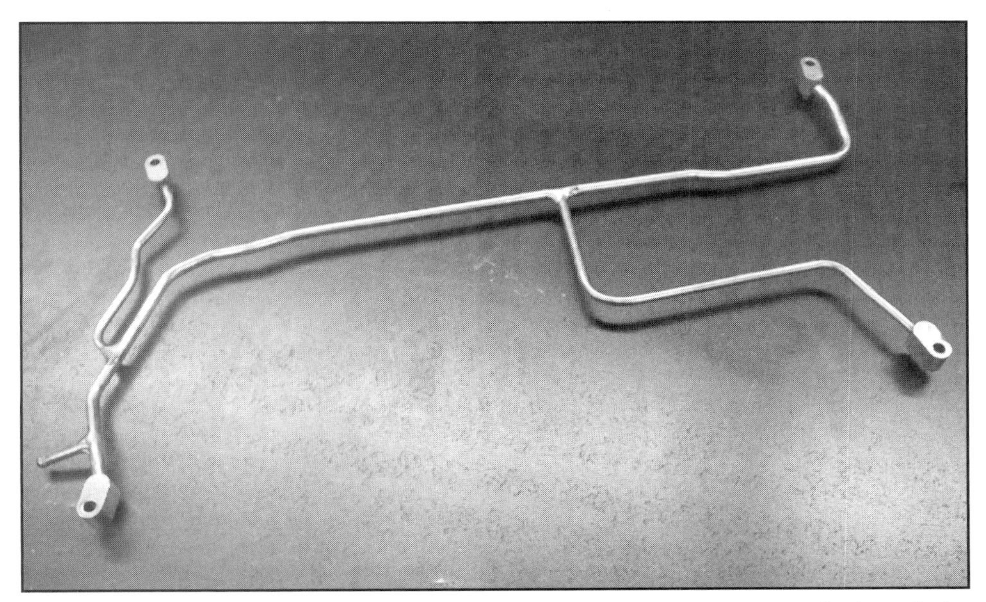

The cylinder head crossovers allow air and vapor in the cooling system to exit the cylinder heads easily.

The plastic tank/aluminum core OEM radiators (shown here with air-conditioning condenser) provide excellent cooling capacity and rarely need to be upgraded.

the cylinder heads to maintain an uphill purge path. This allows air or vapor to exit the cylinder heads easily and move to a pressurized recovery bottle located in the vehicle engine compartment. It is these vent circuit components that must be swapped out when installing an LS6 manifold in place of an LS1 piece. See chapter 2 for specifics on this swap.

The LS1 uses Dex-Cool coolant, which is good for up to 100,000 miles under normal usage. If you wish to improve cooling system performance, you can add a bottle of Red Line Water Wetter or decrease the concentration of Dex-Cool in the system.

How to Modify Your Thermostat for a Cooler Running Engine

This procedure will require that the cooling system be drained to a level below that of the thermostat. With the coolant level adjusted, you can remove the thermostat housing from the front of the engine for disassembly while losing minimal coolant.

Disassembly of the thermostat is quite straightforward: Simply compress the spring and retainer while twisting it out of the two housing retaining posts. You can twist either by hand or with a pair of Channelocks as shown here. It is a good idea to keep your hands on it or even wrap the assembly in a rag, as once the retainer is loose, the spring will do its best to scatter parts all over the room.

With the unit disassembled, note the recessed area inside the housing where the piston seats: this is where the wax motor is located. By placing a small spacer here, you are able to adjust the thermostat to any temperature you desire. But be warned, you can go too far here, preventing the thermostat from closing completely, keeping the engine from reaching proper operating temperature in a timely manner.

A 1/8-inch brass ferrule works well as a spacer because it is easily filed down to the appropriate thickness. Make sure it is a loose fit in the housing to allow room for heat expansion. A spacer of approximately .090-inch thickness makes for a thermostat opening temperature of about 168°. This is the minimum temperature that will allow the thermostat to close completely.

5 Cylinder Heads

The name of the game with cylinder heads is volumetric efficiency. The more efficiently the heads allow air into and out of the cylinders, the more power the engine will produce. General Motors pulled out all of the stops during the LS1 cylinder head development program, and these heads have earned a reputation as some of the best OEM heads ever produced. The LS1 cylinder heads are lightweight, offer good resistance to detonation, and most importantly, flow like gangbusters.

By the Numbers: Composition, Dimensions and Volumes

Before looking at methods commonly applied to modify the LS1 cylinder head, a closer look at the design is in order. The heads have undergone only minor revisions since their introduction in 1997, most notably a switch from perimeter to center-valve cover-bolt configuration for the 1999 model year. All Gen-III cylinder heads are interchangeable.

The Powertrain design team's primary objectives for Gen-III cylinder heads were to achieve identical airflow direction and energy for all cylinders while optimizing fuel injector placement and targeting. To attain this goal, "replicated" ports were designed, where each of the intake ports and each of the exhaust ports are identical in size, shape and volume. The result is perfectly balanced air and fuel distribution between the cylinders, which is the key to maximizing power and enhancing emissions performance.

All Gen-III cylinder heads are sand-cast 356 aluminum (except 1999 and 2000 LQ4, which are cast iron), which is heat-treated to T6 specifications before final machining and assembly. The nominal intake port volume for LS1 passenger car heads is 200cc, the nominal exhaust port volume is 70cc, and combustion chamber nominal volume is 67.3cc. See the nearby table for exact specs of all other Gen-III production heads. The roof of the combustion chamber around the valves blends smoothly with the valve seat's top angle. The valve seat angles are 30°, 45° and 60°. The intake valve size is 2.00 inches, and the exhaust valve size is 1.55

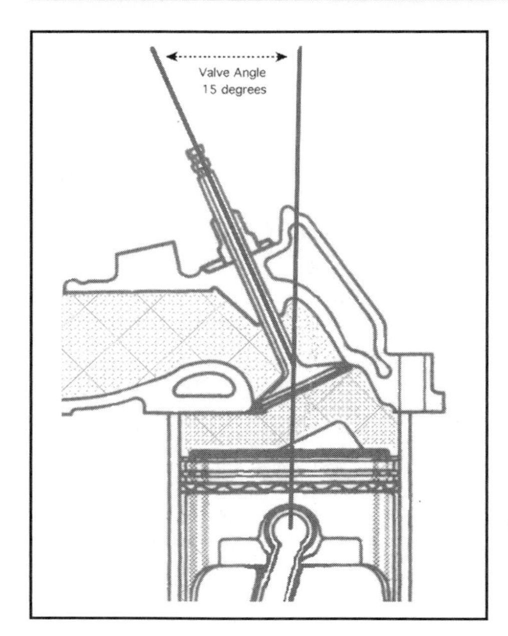

Gen III heads have a shallower valve angle, 15° over the previous 23°, which greatly improves airflow.

Production Gen-III Cylinder Heads at a Glance

Application	GM Part Number	Combustion Chamber Volume	Intake Port Volume	Exhaust Port Volume
1997-1998 LS1	12559853	66.67cc	200cc	70cc
1999-2003 LS1	12564241	66.67cc	200cc	70cc
2001-2003 LS6	12564243	64.45cc	210cc	75cc
1999+ L59, LR4, LM4/LM7	12559852/ 12561706	61.15cc	200cc	70cc
1999-2000 LQ4 (iron)	12561873	71.06cc	210cc	75cc
2001-2003 LQ4 (alum.)	12562317	71.06cc	210cc	75cc
2002-2003 LQ9	12572035	71.06cc	210cc	75cc

inches. Both valves are stainless steel with 8 mm stems. The valve face angles are 30°, 46° and 60°. The valve guides are sintered-iron units and that are pressed in place. Though there are no exhaust gas passages in the Gen-III heads, passages for the coolant vapor ventilation system exist at the front and rear of each cylinder head.

Valve Angle

A critical geometric relationship in any cylinder head is its valve angle. Valve angle is a comparison of the centerlines of the cylinder bore and the valve stem. This dimension plays a role in the shape and volume of the combustion chamber, design and location of the intake and exhaust ports, location of the spark plug, and valve diameter. Speaking in general terms, V8 engines benefit from a head with minimal valve angle. Gen-III cylinder heads have a 15° valve angle, a drastic improvement in airflow over the previous small-block's 23° heads.

The lessened valve angle utilized in Gen-III heads improves valvetrain geometry by lessening the push rod departure angle from the lifter, and improves exhaust flow in the transition between the port floor and the valve seat. This allows a shallow combustion chamber and yields a compression ratio of 10.1:1 with a flat top piston in the passenger car LS1.

The intake port of an LS1 head flows remarkably well in stock configuration.

Intake Port

The key aspect of the LS1 intake port design is that it allows the incoming air an unobstructed path past the intake valve and into the combustion chamber. The ports feature a large runner opening that gradually tapers down so that as the charge air gains speed, it also gains directional stability. Smooth runner-to-valve transition areas keep the air from having to turn right or left to any significant degree. Replicated ports get the air and fuel into the cylinder with the same level of energy from bank-to-bank and port-to-port.

Careful placement of pushrod holes, head-bolt bosses and rocker arm mounting bosses minimize their influence on the shape and location of the intake ports. This also reduced the amount of machining necessary and simplified the assembly process. More space was created for the ports by using four head bolts around each cylinder rather than five as found on previous designs. There is a trade-off for using fewer fasteners though, in the form of lesser clamping capability.

Injector Targeting—Proper fuel injector targeting plays an important role in both idle quality and exhaust emissions. Each port's fuel injector is targeted on the backside of the intake valve to provide ideal fuel vaporization. The intake ports have a notched roof to facilitate the injector.

Combustion Chamber

The combustion chamber volume for an LS1 head is 66.67cc, and its surfaces are sculpted to ease the transition to the valve seat area. The

The LS1 combustion chamber is an efficient, heart-shaped design that diminishes the likelihood of detonation. Combustion chamber volumes range from 61cc for LM4 and LM7 heads, to 71cc for LQ4 and LQ9 pieces.

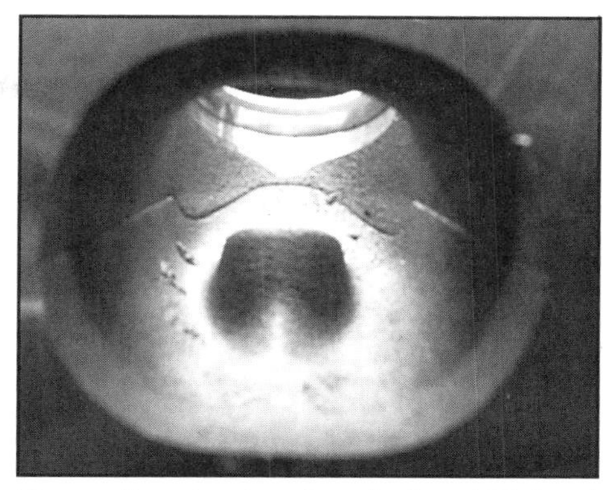

The unaltered exhaust port of an LS1 head provides minimal turbulence to exiting gases. LS6 and LQ9 heads have D-shaped exhaust ports while all others have standard modified oval-shaped ports as shown here.

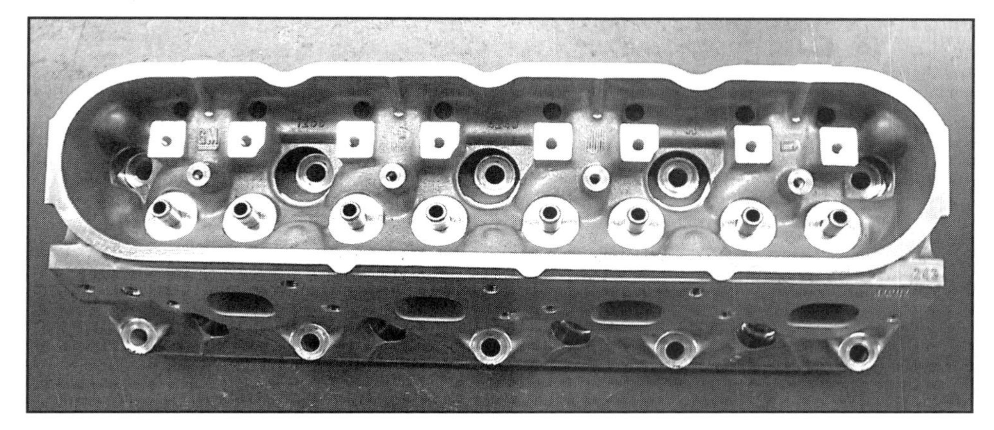

The LS6 cylinder head is completely interchangeable with all other Gen-III heads, and is the best-performing head of the line. Most Gen-III cylinder heads will appear identical to these LS6 heads from above. The exception is the early 97-98 LS1 head, which had perimeter valve cover bolts.

design of the intake port keeps the air fuel mixture moving in a favorable direction with respect to the spark plug, serving to enhance combustion and direct the flame to the center of the bores.

A sharp fillet radius was built into the inlet side of the combustion chamber to induce swirl into the incoming air, and in turn concentrate mixture motion around the spark plug boss area. The spark plug is positioned near the center of the chamber allowing the flame to travel in its natural spherical pattern, equally in all directions, thereby reducing the likelihood of quenching. Flame propagation is enhanced by matching the radial chamber size to the bore diameter, thereby decreasing crevice volume.

Exhaust Port

A good cylinder head must allow exhaust gas out as freely as it allows the charge air in. Gen-II heads and most other two valve heads have a turbulence and flow problem in the area where the port floor transitions to the valve seat. This area is commonly known as the short side radius. The open design of the Gen-III combustion chamber combined with the 15° valve angle rids the short side radius of many of these problems.

LS6 Heads

The LS6 cylinder head is a further refined version of the LS1 head, meaning differences between it and the LS1 head are relatively few. It has an enlarged intake port volume of 210cc, and at almost 65cc, the combustion chamber is slightly smaller and more efficient than the LS1. This more efficient design raises compression to 10.5:1, shortens burn times, and ultimately means less ignition timing advance is required to produce the same power. And because less advance allows more efficient combustion, the LS6 heads allow the engine to produce more torque. The exhaust port volume was increased to 75cc and had a unique D-shape that improves flow.

Truck Heads

There are several Gen-III truck

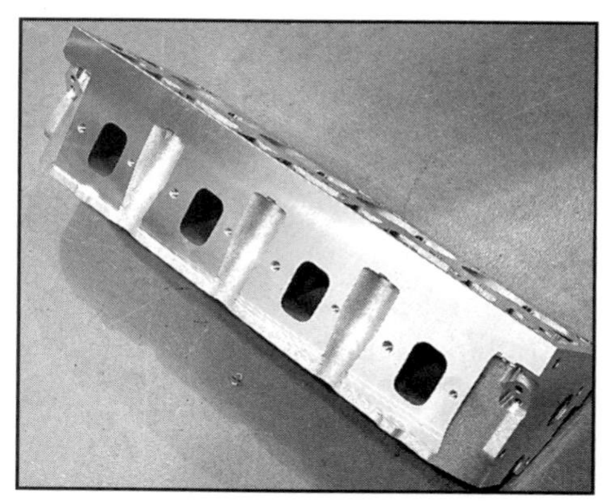
C5R heads incorporate rectangular intake ports designed specifically for Kinsler intake manifold. OEM manifolds are not compatible.

cylinder head variants of particular interest to the high-performance enthusiast. The LQ4 and LQ9 heads are found on 6.0-liter truck engines and offer larger combustion chambers that lower compression ratio, making them perfect for a forced induction application. Particularly good is the LQ9 head, which debuted on the Cadillac Escalade in 2002. This head is virtually identical to the LS6 head, with the same improved intake and exhaust ports, but has a larger, 72cc combustion chamber for lower compression than the LS6 variant. Swapping on the LQ4 or LQ9 head drops the compression ratio of a typical LS1 or LS6 engine to approximately 9.5:1.

The LR4 and LM7 heads found on 4.8- and 5.3-liter truck engines offer smaller combustion chambers, which will increase compression ratio, but otherwise have no advantage over an LS1 head. The smaller chambers allow a slightly higher compression ratio to be achieved over other heads. Milling these heads .030-inches will bring compression to nearly 11.5:1 on a stock LS1 shortblock. Though the extra compression can be valuable in a race engine, be careful, as detonation can become an issue at compression ratios over 11:1. This is particularly true when using 92-octane pump gas. Use of a camshaft with a lot of overlap can help, because it will bleed off some cylinder pressure at lower rpm. Because these heads are equipped with small, 1.89-inch intake valves, a valve upgrade is practically mandatory.

Spotter's Guide to Cylinder Head Casting Numbers

Casting numbers are a sure way to determine the origin of any Gen-III cylinder head.

An easy way to identify most Gen-III cylinder heads on an assembled engine is by the casting number in the lower left- or lower right-hand corner of the head, just beneath the valve cover rail. The three-digit number correlates to the last three numbers of the GM part number. The only known exception is the iron LQ4 heads, which do not exhibit an external casting number, but their ferrous composition (and rust) should make them easy to recognize. In any case, the complete part number usually appears between the pushrod holes as shown above.

All cylinder heads in the Gen-III family are interchangeable, but when swapping heads you should be aware of which years you are dealing with. The reason for this is valve-cover compatibility. Covers from '97 and '98 use perimeter fasteners, while '99 and later use center bolt-style covers and updated coil pack brackets. All other features and dimensions are the same, meaning you can swap intakes, exhaust manifolds or headers to your heart's content.

C5-R Heads

Because of their high cost, GM Performance Parts C5-R heads are jokingly believed to be cast from commercially pure unobtainium, but in fact; they are the same 356 aluminum as the production heads. The heads were developed for GT2 endurance racing at events like the 24 Heures du Le Mans and the 24 Hours of Daytona. They are similar in appearance to the production Gen-III heads, but have several important differences. Notably, the valve angle was revised to 11° in order to improve airflow and create a smaller combustion chamber. The inlet port floor and roof are raised to accommodate a Kinsler intake manifold, so production manifolds will not fit. The valve spring seats are machined to accept 1.630-inch springs and a receiver groove is machined into the valve cover rail for an O-ring seal. There are three versions of the C5-R head offered in various stages of machine work. You can see the specs of these and the other factory heads later in the chapter.

Choosing the Right Head

As stated earlier, LS1, LS6, LQ4

Head Specs & Swaps

GM has allowed us incredible flexibility in swapping heads by making all Gen-III cylinder heads interchangeable. But did you realize: to date, they have been offered in no fewer than seven varieties from the factory? And when you add in GM Performance Parts' offerings, the total is approaching a dozen. While having choices is great, many are understandably confused by the dizzying array of options. The following data has been compiled from various sources inside GM. Chances are, there is a head that fits your performance goal perfectly.

1997+ LS1 5.7-liter Passenger Car

Material:	Aluminum
Part Number:	12559853 (1997–98)
	12559853 (1999–00)
	12564241 (2001–03)
Combustion Chamber Vol.:	66.67cc
Compression Ratio:	10.1:1
Intake Port Vol.:	200cc
Exhaust Port Vol.:	70cc
Intake Valve Dia.:	2.00 inches
Exhaust Valve Dia.:	1.55 inches

What you need to know: The standard issue LS1 head is the best all-around head for street or street/strip engines. A thorough porting and milling job plus a valve upgrade on these will really wake up an engine. Milling the deck of these heads .030 will yield a compression ratio of approximately 10.9:1 on a 5.7-liter LS1. The heads have undergone only minor revisions since their introduction in 1997, most notably a switch from perimeter to center valve cover bolt configuration for the 1999 model year. Each style has dedicated valve covers and coil pack mounting apparatus.

2001+ LS6 5.7-liter Passenger Car

Material:	Aluminum
Part Number:	12564243
Combustion Chamber Vol.:	64.45cc
Compression Ratio:	10.5:1
Intake Port Vol.:	210cc
Exhaust Port Vol.:	75cc
Intake Valve Dia.:	2.00-inches
Exhaust Valve Dia:	1.55-inches

What you need to know: The LS6 cylinder head is essentially a tuned-up LS1 head. At 65cc, the combustion chamber is slightly smaller and more efficient than the LS1. This more efficient design increases compression to 10.5:1, shortens burn times, and ultimately means less ignition timing advance is required to produce the same power. And because less timing allows more efficient combustion, the LS6 heads allow the engine to produce more torque. The exhaust port is a unique D-shape that improves flow. Milling the deck of these heads .030 inches will yield a compression ratio of approximately 11.1:1 on a 5.7-liter LS1. LS6 heads are the best choice only when all-out power is needed. Be prepared for a hefty price tag at the dealer or steep core charge from your head porter.

1999+ LR4 4.8-liter Truck
1999+ LM4 / LM7 5.3-liter Truck

Material: Aluminum
Part Number: 12559852, 12561706
Combustion
Chamber Volume: 61.15cc
Compression Ratio: 9.5:1
Intake Port Vol.: 200cc
Exhaust Port Vol.: 70cc
Intake Valve Dia.: 1.89-inches
Exhaust Valve Dia.: 1.55-inches

What you need to know: These truck heads offer no advantage over an LS1 head except the smaller combustion chamber. This, along with milling of the deck surface will allow a slightly higher compression ratio to be achieved. Unmodified, they will yield a compression ratio of approximately 10.9:1, milling the deck .030 inches will yield a compression ratio of approximately 11.5:1 on a 5.7-liter LS1. Because of the small intake valves installed in these heads, a valve upgrade is practically mandatory.

1999–2000 LQ4 6.0-liter Truck

Material: Cast Iron
Part Number: 12561873
Combustion
Chamber Volume: 71.06cc
Compression Ratio: 9.5:1
Intake Port Vol.: 210cc
Exhaust Port Vol.: 75cc
Intake Valve Dia.: 2.00-inches
Exhaust Valve Dia.: 1.55-inches

2001+ LQ4 6.0-liter Truck

Material: Aluminum
Part Number: 12562317
Combustion
Chamber Volume: 71.06cc
Compression Ratio: 9.4:1
Intake Port Vol.: 210cc
Exhaust Port Vol.: 75cc
Intake Valve Dia.: 2.00-inches
Exhaust Valve Dia.: 1.55-inches

What you need to know: The LQ4 heads are found on the low-performance versions of the 6.0-liter truck engine. They offer a larger combustion chamber that lowers compression ratio, making them perfect for a forced induction application. The LQ4 head casting was changed to aluminum starting in model year 2001. Swapping on unmodified LQ4 head drops the compression ratio of a 5.7 liter LS1 engine to approximately 9:1. This is the workhorse head for street/strip turbo and blower cars.

2002+ LQ9 6.0-liter Truck

Material: Aluminum
Part Number: 12572035
Combustion
Chamber Volume: 71.06cc
Compression Ratio: 10:1
Intake Port Vol.: 210cc
Exhaust Port Vol.: 75cc
Intake Valve Dia.: 2.00-inches
Exhaust Valve Dia.: 1.55-inches

What you need to know: The LQ9 heads are found on the high-performance versions of the 6-liter truck engine, and debuted in the Cadillac Escalade for 2001. This head is virtually identical to the LS6 head, with the same improved intake and exhaust ports, but like the LQ4 head, has a large 72cc combustion chamber. This is the head of choice for max-effort forced induction engines.

GMPP "Gen-III Racing"

Material: Aluminum
Part Number: 12480176
Combustion
Chamber Volume: 65cc
Compression Ratio: 10.5:1
Intake Port Vol.: 210cc
Exhaust Port Vol.: 65cc
Intake Valve Dia.: 2.00-inches
Exhaust Valve Dia.: 1.55-inches

What you need to know: GM Performance Parts offers these fully assembled, CNC-ported LS6 cylinder heads with stock size valves and a basic "clean up" of the ports.

GMPP "Showroom Stock Racing"

Material: Aluminum
Part Number: 25534321
Combustion
Chamber Vol.: 63cc
Compression Ratio: 10.8:1
Intake Port Vol.: 200cc
Exhaust Port Vol.: 70cc
Intake Valve Dia.: 2.00-inches
Exhaust Valve Dia.: 1.55-inches

What you need to know: This is another GM Performance Parts assembled LS1 head that has been milled to bring combustion chamber vol. down to 63cc. Compression ratio is increased to nearly 11:1. No other alterations are made to the castings.

GMPP C5-R

Material:	Aluminum
Part Number:	12480005
Combustion Chamber Vol.:	38cc
Compression Ratio:	N/A
Intake Port Vol.:	00cc
Exhaust Port Vol.:	70cc
Intake Valve Dia.:	N/A
Exhaust Valve Dia.:	N/A

GMPP C5-R

Material:	Aluminum
Part Number:	12480025
Combustion Chamber Vol.:	38cc
Compression Ratio:	N/A
Intake Port Vol.:	00cc
Exhaust Port Vol.:	70cc
Intake Valve Dia.:	N/A
Exhaust Valve Dia.:	N/A

GMPP C5-R

Material:	Aluminum
Part Number:	12480090
Combustion Chamber Vol.:	30cc
Compression Ratio:	N/A
Intake Port Vol.:	00cc
Exhaust Port Vol.:	70cc
Intake Valve Dia.:	N/A
Exhaust Valve Dia.:	N/A

What you need to know: The GM Performance Parts C5-R heads are similar in appearance to the production Gen-III heads, but have several important differences. Notably, the valve angle was revised to 11° in order to produce a smaller combustion chamber. The inlet port floor and roof are raised to accommodate a Kinsler intake manifold, so production manifolds will not fit. The valve spring seats are machined to accept 1.630-inch springs and a receiver groove is machined into the valve cover rail for an O-ring seal. There are three versions of the C5-R head offered in various stages of machine work, but all three versions will require significant machine work before assembly.

and LQ9 cylinder heads are completely interchangeable, and each has its own merits. So the key question is, which heads are right for your engine?

The chart above indicates that the LS6 heads have a definite advantage over the LS1s. This may lead you to believe that since you can pick up a solid 35cfm with a pair of stock LS6 heads, they may be a cost-effective alternative to porting of the stock pieces. While it sounds good in theory, the reality is that LS6 castings are expensive, and even a conservatively ported pair of LS1 heads will out-flow the stock LS6's by a substantial margin. If you do choose to go with the LS6 heads, you can expect something in the neighborhood of a 10rwhp increase over stock or similarly ported LS1 heads on a 346-inch engine. Don't forget to factor in the intake manifold, because as the table on this page shows, it will significantly impact power numbers.

Recommendations

If you are building a street/strip car, ported LS1 heads offer the best bang for the buck. It is true that LS6 heads will make more power, but the extra expense of the castings does not make them a worthwhile expenditure for anything but a dedicated racecar. On the other hand, if you already have a pair of LS6 castings, by all means use them.

Aluminum LQ4 or LQ9 cylinder heads are most beneficial on an engine that will use a turbo or supercharger. The drop in compression ratio provided by the increased combustion chamber volume is a big advantage on a forced induction mill, particularly when the stock shortblock will be used. Even when building a complete engine, lowering the compression ratio by using a head with a larger chamber is preferable to installing lower compression pistons, particularly in an LS1. Because the LS1 piston is very short to begin with, further lowering the compression

Comparison of Unmodified Cylinder Heads Flow in CFM

As the table below shows, the difference in intake airflow between the LS1 and LS6 cylinder heads is substantial when comparing stock unmodified units. The LS6 is clearly superior here, out-flowing the LS1 heads by more than 30cfm at .500-inch lift. But the numbers don't tell the whole story

Cylinder Head	Valve Lift in Inches							
	.200	.300	.350	.400	.450	.500	.550	.600
LS1	137	187	207	223	228	237	242	243
LS6	156	204	225	243	257	268	275	278

Note: Both heads measured at 28" Depression

Dyno Testing LS1 vs. LS6

To show the difference between the castings, I tested an engine in a 1998 Z28 on a chassis dynamometer with modified LS1 and LS6 heads and intake manifolds. I mixed and matched to determine relative power numbers from the various combinations. Both pairs of heads were thoroughly ported by Agostino Racing Engines and upgraded to Manley 2.055 intake and 1.60 exhaust valves. The same cam (Lunati 222°/230°duration at .050-inches lift, .534/.544 inches of lift with a 114-degree lobe separation angle) was used throughout. Other components of note: Grotyohann 1 3/4-inch headers, stock injectors, ported MAF, and 28° of timing. The car was a six-speed manual with 4.11 gears in a 12-bolt rearend. This is not a statement of maximum power available from these heads or similar combinations; it is for comparison only.

Cylinder Head	Intake Manifold	Horsepower	Torque
Ported LS1	LS1	406	393
Ported LS1	LS6	423	405
Ported LS6	LS1	413	400
Ported LS6	LS6	435	416

Properly ported cylinder heads are equal parts art and science.

height compromises their integrity. Weak pistons are not a desirable trait, especially in a forced induction engine that is more likely to be subjected to detonation. As with the LS6 heads, the LQ9 castings are rather expensive. If you want to maximize your bang for the buck, stick with the 2001 and later aluminum LQ4 castings.

Inspection

Regardless of which heads you select, it is an excellent idea to make sure the heads you will be working on or sending out are in good condition. Nothing is worse than starting (or finishing) extensive porting on a head only to find out the casting is damaged. Spend the few bucks necessary to have the heads thoroughly inspected for cracks. Cracked Gen-III heads are not common, but it is not unheard of either so this is cheap insurance. It is also a good idea to pressure test the heads to check for cracks or porosity problems in non-visible areas such as the water jackets. While TIG welding is an option to repair a damaged head, you are much better off starting with new pieces.

Don't make the mistake of purchasing a pair of ported cylinder heads based on dimensions and flow numbers alone. Bigger is not always better. Massive ported runners and big valves look impressive, but are no guarantee that a pair of heads will work for your combination. Like anything else, much thought and planning must be dedicated to a cylinder head package before modifications and components are selected. How is the car to be used? What is the engine's displacement? Desired rpm range? Camshaft to be used? What other modifications have been done? Will there be a power adder in use? What is the ultimate power goal for the car? Clearly, there is much to discuss when addressing the topic of cylinder heads, so you need to consider all of your variables..

Modifying Cylinder Heads

Modifying an engine's cylinder heads is one of the best ways to increase its power. However, high-quality head work is also amongst the most time-consuming, and it is also fairly difficult, so it can be very expensive. If you're buying ported heads, you want to be sure you get the best match for your engine the first time. Because there are currently no aftermarket cylinder heads available for the Gen-III engine family, porting of OEM heads is necessary.

Stage What?

There are a few terms you will hear being thrown around with regularity when speaking of ported heads. Stage 1, 2 and 3 are common labels applied to varying levels of head modification. Every shop has its own way of labeling its products, so it is up to you to make sure you ask the right questions to get what you want. The most important thing is to not get hung up on Stage this or Stage that.

Stage 1—Stage 1 heads usually consist of a basic port job, upgraded 2.00 or 2.02 intake/1.57 exhaust valves and upgraded springs. These are best when combined with a mild cam in the 218° of duration and .550-inches of lift range, and will generally make about 380 rear-wheel horsepower.

Stage 2—Moving to a Stage 2 head usually denotes a more thorough porting job, 2.055 or 2.08 intake/1.60 exhaust valves, upgraded springs, and maybe some trick parts like titanium spring retainers. Stage 2–style heads generally respond very well to cams

Flow benches such as this Superflow SF-600 should be part of every serious head porter's arsenal.

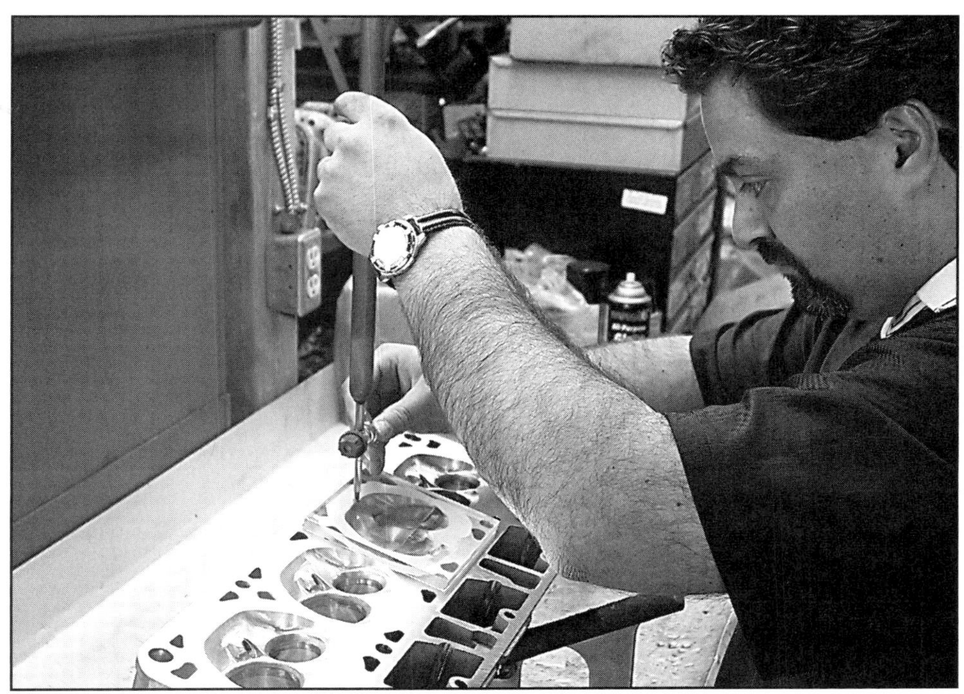

A cylinder head's various volumes are measured using a burette as shown here. This is done to ensure consistent volumes, and thus compression ratios, between cylinders.

with 225° to 230° of duration and up to .600-inches of lift. Expect 410 to 425 rear wheel horsepower.

Stage 3—These heads usually expand on the items of a Stage 2 setup. At this level, intake valve sizes can increase to 2.10 or 2.125 and exhaust valves can be as large as 1.625, valve shrouding becomes an issue with valves this large on stock-bore engines. Intake and exhaust port volumes are usually larger; making these heads best for very big stroker engines.

Flow Benches

A flow bench measures cylinder head flow. Flow benches are precision pieces of equipment, but they can be manipulated to produce nearly any result desired by the operator. It is possible to get quite different numbers from the same head between two benches or even two operators on the same bench. The only sure way of accurately comparing flow numbers between cylinder heads is to test them on the same bench with the same guy working the controls. The upshot of this is that unless you plan to race a flow bench, reserve judgment on a heads' flow numbers until you have all the facts. The real test comes when the final product is strapped to the dyno or run down the track.

While cylinder head flow numbers are informative, they don't tell the whole story of how a head will perform. Flow numbers must be correlated to port volume because as the volume of the port increases, so too should flow.

Intake Port Volume

Most people expect to see intake ports the size of the Holland Tunnel on a pair of ported heads. The truth is that extensive modifications in this area often end up being counter-productive, particularly on a smaller displacement engine. Opening up the intake ports obviously increases port volume and it is port volume that establishes the relative potential for airflow. But when the ports are too large, the charge velocity slows and kills low-end torque.

Port Volume and Its Effect on Velocity

Big ports on small engines make for slow cars. This is true for street engines that spend the majority of their time below 6000 rpm; all-out race engines are another matter. The key to producing torque in any engine is intake port velocity. If the intake port is too large, velocity falls, sacrificing torque. However, more port volume is required to produce

Cylinder Head and Manifold Flow Testing

I have worked with Agostino Racing Engines in Toronto in performing extensive airflow testing on every conceivable combination of stock and ported LS1 and LS6 cylinder heads, separately and with intake manifolds. All ported heads were equipped with Manley 2.055/1.60-inch valves. The tests provided some interesting results.

Of particular note is the fact that the ported LS1 head / LS6 manifold combination outflows the ported LS6 head / LS6 manifold combination until valve lift reaches .450-inches. At that point, the LS6 / LS6 combination overcomes, but think about it: When open, where does the valve spend most of its time? If you said at lower lift, you are on the right track.

Also worth noting is the substantial airflow gain shown by the LQ4 truck manifold over the LS1 piece, especially at low valve lift. Again, the margin narrows at higher lift. One other important note about the truck manifold: It won't fit beneath the F-body's cowl without modification, so don't knock yourself out searching the bone yards for one.

It is interesting that when the heads are ported, the margin in airflow between them closes significantly. Specifically, airflow is within 10-cfm throughout the tested range. If you study it carefully, you will see that the ported LS1 heads actually out-flow the ported LS6 heads by a small margin until .350-inch valve lift. From there on, the LS6 has the edge, but it is a narrow one. I believe the airflow numbers below validate my opinion that the ported LS1 heads with the LS6 manifold represent the best combination for the dollar.

Cylinder Head and Intake Manifold Airflow Comparison
(in cubic feet per minute-cfm)

Cylinder	Intake	Valve Lift							
Head	Manifold	.200"	.300"	.350"	.400"	.450"	.500"	.550"	.600"
Stock LS1	Bare Head	137	187	207	223	228	237	242	243
Stock LS6	Bare Head	156	204	225	243	257	268	275	278
Ported LS1	Bare Head	164	217	234	253	270	284	296	305
Ported LS6	Bare Head	159	212	236	255	272	300	307	314
Stock LS1	LS1	136	184	200	214	222	227	229	235
Stock LS1	LS6	136	186	206	223	227	236	241	242
Stock LS6	LS1	156	199	212	224	232	238	243	247
Stock LS6	LS6	154	204	220	235	247	257	263	265
Ported LS1	LS1	141	185	205	222	232	241	251	258
Ported LS1	LQ4	163	206	224	237	249	262	265	270
Ported LS1	LS6	166	211	229	244	257	269	277	283
Ported LS6	LS1	156	199	217	230	242	253	263	268
Ported LS6	LS6	153	202	222	241	257	270	280	289

good power at extremely high rpm or on large displacement engines. But it is more complicated than that, if the cross-sectional area of the port is increased, flow will usually, but not always, increase too. What does this mean to you?

Simply this: minimize port cross-sectional area throughout the runner to a point that still allows the engine to attain target rpm. This is accomplished by removing only as much material from the port walls as necessary to achieve your target volume and flow numbers. Naturally, there is a formula that prescribes the amount of minimum intake runner cross-sectional area that an engine requires. It looks like this:

(Cylinder Volume x Peak Torque rpm) / 88,200 = Minimum Intake Port Area

So, if we are solving for a stock displacement, 346 ci dLS1, with a peak torque target of 5000 rpm, we would have this:

Example: [(346 / 8) X 5000] / 88,200 = 2.452 square inches

What you should take from this is that small intake ports generally tend

Porting of the intake runners should focus on maximizing flow velocity by minimizing cross-sectional area, as illustrated by the cylinder head on the right.

Exhaust porting should be done to complement the amount of work done in the intake ports. A dual-pattern camshaft can help compensate for undersized exhaust ports.

to make an engine very responsive because they are efficient at filling the cylinders at lower engine speeds. Large ports are far better at generating efficient flow velocities at high engine speeds where the small ports become an airflow restriction.

The 200cc intake port volume in a pair of stock LS1 heads is sufficient to support a 346 cubic-inch engine to over 6600 rpm. By contrast, the same head would run out of steam by 6000 rpm on a big 422-ci mill. Conversely, a head with 220cc runners would be perfect for a 422 that will regularly see 6500 rpm, but would be really soft on a 346-ci engine.

Clearly, high-intake port velocity is a good thing. Be warned though, as you can have too much of that good thing: A condition known as *intake port stall* occurs when intake charge velocity reaches 1200 fps. When this occurs, the intake charge separates, meaning the fuel precipitates out of the air stream and causes puddles in the intake tract. This results in a dramatic efficiency loss, so much so that it acts as a sort of artificial rev-limiter, not allowing the engine to make power at higher rpm. Turbulence in the runner causes many of the same problems as port stall, and negatively impacts power.

Exhaust Ports

Though the intake port gets much of the attention, an efficient exhaust port is just as important for making power. A cylinder head that has excellent intake ports with high velocity will probably make good torque, but unless the head is teamed with an equally good exhaust port, it won't generate maximum horsepower. This is because at higher engine speeds, the exhaust port will not have time to evacuate all of the residual exhaust gases from the chamber before the next cycle starts. The exhaust ports should therefore be opened up proportionately to the intake ports.

E/I Ratio—An excellent method to evaluate a cylinder head is to apply a simple equation known as the Exhaust/Intake (E/I) ratio. To use this formula, exhaust port flow is expressed as a percentage of the intake port flow at a specific valve lift. Most engine builders prefer an exhaust/intake ratio of 75–85%.

For example, take a head that has an intake port that flows 300 cfm and an exhaust port that flows 240 cfm, both at .500 valve lift.

Example: 240 / 300 = .80

Dividing the exhaust flow of 240 cfm by the intake flow of 300 cfm produces an 80% exhaust/intake ratio. Because this imaginary head has respectable intake flow numbers and a near-ideal E/I ratio, it is likely to make good power.

Another important aspect of the E/I relationship has to do with valve timing. A split duration camshaft (with a longer duration exhaust lobe) can help compensate for a head with a less than ideal exhaust port. A cylinder head with an optimized exhaust/intake ratio is usually better off with a single pattern cam. For more information on camshafts, see chapter 6.

Valve Bowls

The bowl is arguably the most important area in any cylinder head.

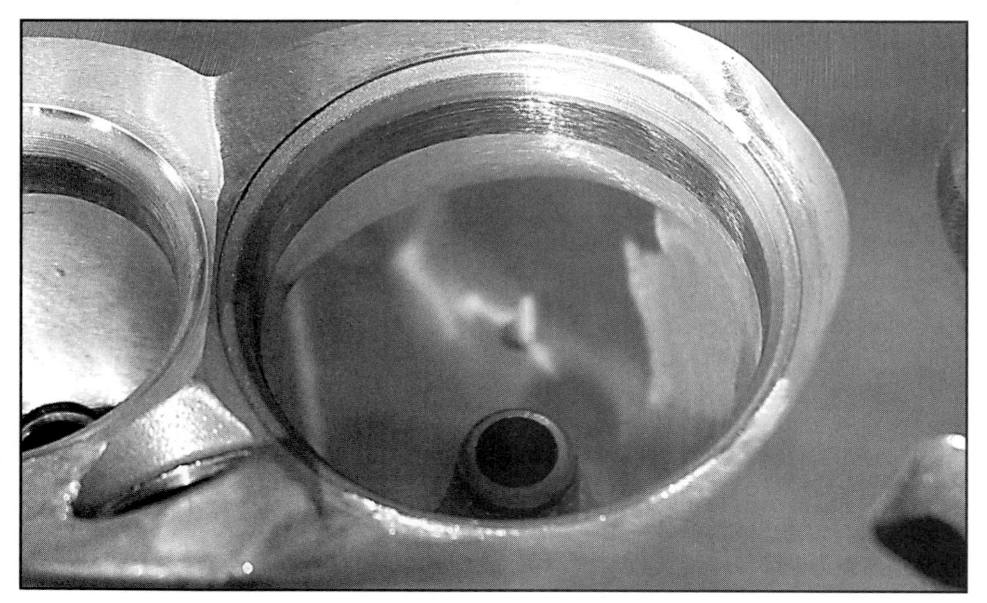

The valve bowls should be cut to 89–90% of valve diameter and blended smoothly with the runners.

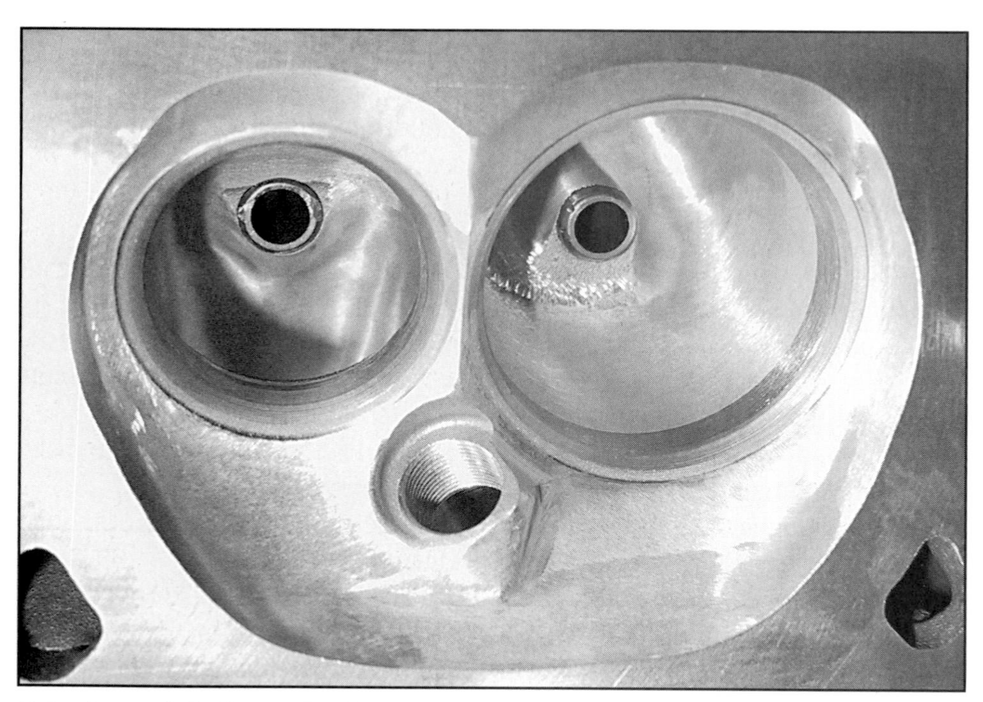

Extensive work in the combustion chambers is not recommended for most engines. Mild smoothing of the chamber is all that is needed.

Bowl shape and size are critical to smooth airflow, and become even more significant when larger valves are to be installed. The diameter of the valve bowl diameter should be 89-90% of the valve size. This opens a more direct flow path to the valve and creates a Venturi effect, straightening the charge's movement past the valve and into the combustion chamber. All contours in and around the bowl area should be smoothed and blended, concentrating specifically on the port to valve seat transition. These surfaces should be shaped into smooth radii, while removing as little material as possible.

Combustion Chamber Modifications

The LS1 combustion chamber is an extremely efficient, heart-shaped chamber that needs little modification for most applications. Combustion-chamber shape affects volumetric efficiency by its influence on valve size and degree of swirl. GM designed a high degree of swirl into the chambers, and there is seldom any need to mess with it. Extensive mods performed to improve the chamber will lower the compression ratio and usually net zero gain.

Fast-Burn Chambers

Fast combustion is highly desirable in an internal combustion engine because the fuel burn time occupies a shorter portion of the crank angle interval at a given engine speed than a slow-burning combustion chamber, increasing efficiency. This not only increases efficiency, but also gives the engine much better tolerance to a lean-mixture and introduction of EGR gases. In addition, a fast-burning chamber yields improved fuel economy due to reduced pumping work and lower heat transfer.

There are several ways in which to control the speed of combustion. One method is to vary the location of the spark plug in the combustion chamber. The location of the spark plug plays a role in the size of the flame front surface area. The flame front area is the leading edge of the spherical surface of the flame contained within the combustion chamber. The larger this surface, the faster the fuel can be burned.

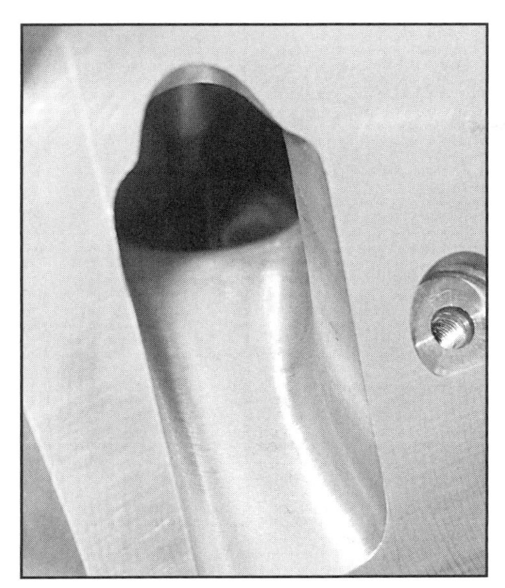

Ported intake runners should allow increased airflow with no degradation in velocity. This means you should smooth the port walls, floor, and ceiling, but remove only as much material as necessary to achieve a smooth, regular surface.

Turbulence

One of the most important factors affecting the burn rate (and thus, efficiency) in the combustion chamber is the turbulence level of the charge at both the time of ignition and during the burn. Though turbulence in the intake port is bad, it is quite beneficial to the combustion process in the combustion chamber.

The shape of the combustion chamber contributes to turbulence by creating a squish area, which is the area where the piston crown comes very close to the cylinder head. The piston crown's close proximity to the cylinder head creates a "squishing" of the air-fuel mixture and pushes it into the non-squish area, thereby increasing the tumble flow in the cylinder. Tumble is rotational turbulence perpendicular to the axis of the cylinder.

Small-scale turbulence generated during the intake stroke, while impacting air/fuel mixture, does little to affect burn duration. Small-scale vortices rotate against each other, but quickly dissipate during the compression stroke. Large-scale motion, on the other hand, has been shown to persist to the time of ignition and greatly increases the burn rate.

For street heads simply rid the exhaust port of any casting flash and surface irregularities.

Swirl is rotational flow in which the axis of rotation is parallel to the axis of the cylinder. Generating rotational swirl from a port is quite easy. However, attaining high flow with the desired swirl rate is much more difficult. The high degree of swirl internal to the Gen-III can be partially attributed to a deflector located near the spark plug.

Some head porters reshape the combustion chambers by welding in additional material to increase the compression ratio. While this method works well in order to maintain valvetrain geometry and valve to piston clearance, it is an expensive and unnecessary alternative that oftentimes adversely affects swirl. If an increase in compression ratio is desired, milling the head deck is a much better alternative. It is best to perform only minor work in the chambers, optimizing airflow by unshrouding the valves.

Head Porting

Basic cylinder head porting is not beyond the abilities of the average do-it-yourselfer, but a bit of discretion with the grinder goes a long way. Only minor work to the intake port entry area and runner is necessary for most 346 cid engines. There is little reason to open the runners up dramatically for a typical stock short block engine. It is good practice to gasket match the intake and exhaust ports at the flanges, but taper them smoothly to the existing port walls. The walls should be smoothed and removed of any irregularities.

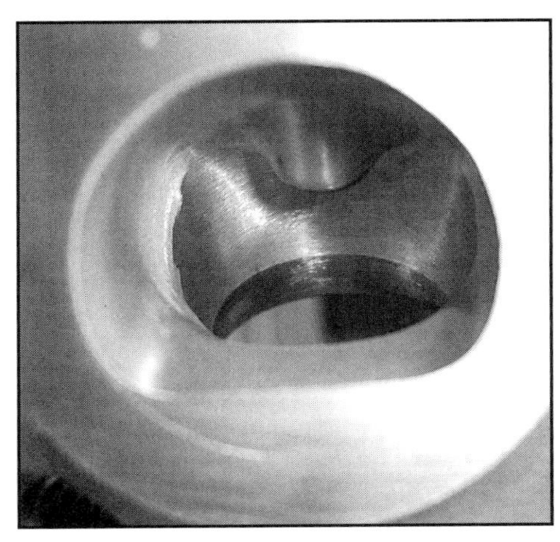

The valve bowl must allow a smooth transition between the combustion chambers and the runners.

Maximum valve size is ultimately restricted by cylinder bore diameter.

Milling

After the port work is complete, the heads should be milled. Milling is a procedure that "trues" the deck of a used cylinder head. This is crucial to achieve proper head gasket sealing. It can also serve to decrease combustion chamber volume. Many shops mill the cylinder head deck 0.030-inches in order to increase the compression ratio. With an LS1 head, this raises the compression to approximately 10.9:1 on a stock 346 cubic inch engine. This is a good practice, as it produces a perfectly streetable compression ratio that requires only 93-octane fuel.

A similar cut on an LR4/LM7 head will result in a compression ratio approaching 11.5:1. In most cases, this is too much for pump gas. For applications where high octane race fuel will be used, it is possible to mill as much as .060-inches from LS1 heads, resulting in a compression ratio of approximately 11.6:1. Be aware that milling the heads will require a corresponding decrease in push rod length.

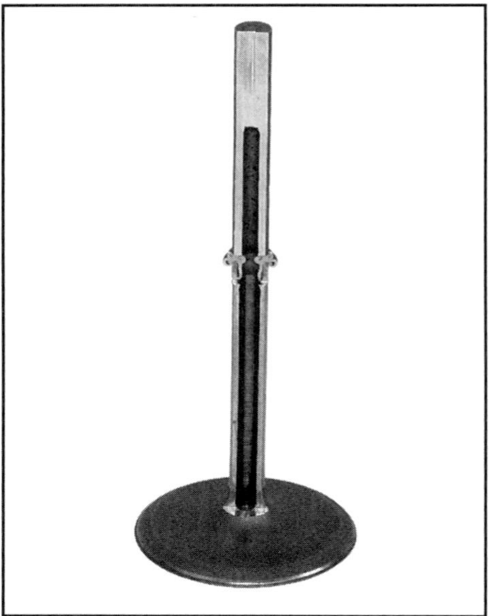

Sodium-filled exhaust valves were introduced to the Gen-III in the 2002 LS6. Sodium filled valves offer improved heat dissipation over standard stainless steel valves.

Valves

As with many things, there is a point of diminishing returns when increasing cylinder head valve diameter. Valve size is effectively limited by combustion chamber shrouding, and ultimately, cylinder bore size. Installing larger valves without the appropriate porting modifications to increase airflow is a waste of time and money. There will be little, if any, power gain from larger valves if the cylinder heads are incapable of providing additional airflow.

Quality valves are available from several companies, the most popular being Ferrea, Manley and Racing Engine Valves (REV). Regardless of the valve brand or size used, valves with a 30° back cut will improve low lift flow.

Installation of larger valves in Gen-III cylinder heads is a great way to improve airflow, assuming proper porting is in place. Stock LS1 valves are 2.00- and 1.55-inches in diameter.

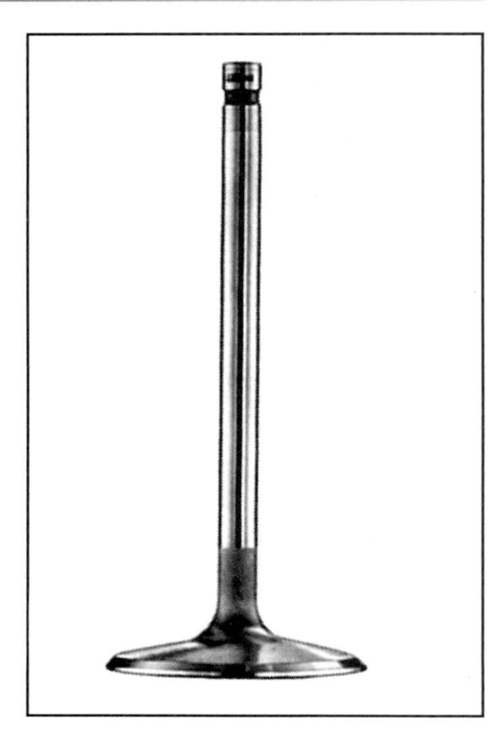

Manley valves are a great choice for high-performance LS1's, as they offer a 30° back-cut right out of the box.

Commonly used performance intake valve sizes for 346 cubic-inch engines are 2.02- or 2.055-inches, and are usually matched with exhaust valves of 1.57- or 1.60-inches. Valves with diameters of 2.02-inches and 1.60-inches or less do not require replacement of the valve seats with larger diameter pieces, and thus are popular for budget-oriented builds. Valves larger than 2.02 and 1.60 do require valve seat replacement, and therefore usually significantly increase the costs involved. Larger LS1's typically receive 2.08/1.60-inch valves. It is possible to outfit Gen-III heads with 2.125/1.625-inch valves by moving the valve guides and seats apart. This is a labor-intensive chore best performed by a machine shop with extensive cylinder head experience.

Valve Material

Most aftermarket valves are stainless steel, and for good reason.

Bronze valve guides are durable and if installed with proper valve stem clearance, not likely to leak.

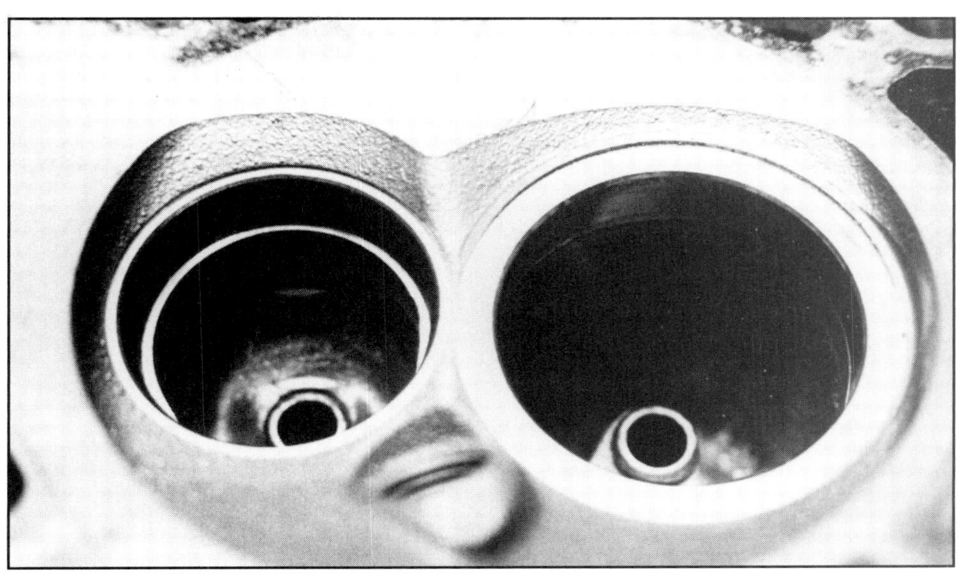

In order to utilize valves larger than 2.02/1.60-inches, larger valve seats must be installed in the head.

Stainless is the best material for the job in most applications. While there are valves available in lightweight titanium or Inconel, they are beyond the scope of this book.

Valve Guides

Valve guides are an essential component of the cylinder heads because they position the valve correctly in the valve seat. A loose valve guide will allow the valve to move around and strike the valve seat in a different place each time the valve closes. This destroys the precise angles machined into the seat, quickly ruining an otherwise good valve job, killing airflow and subsequently sacrificing power. A loose valve guide will also allow oil to leak past the valve guide seal. This oil contaminates the incoming air/fuel mixture resulting in power loss and creating valve stem deposits that obstruct air flow.

Bronze valve guides are the best choice for the majority of LS1 applications. Valve guide-to-stem clearance should be .0010 to .0015-inches for intake and .0015 to .0020-inches for the exhaust. The additional clearance for the exhaust valve is required to allow for expansion as it is obviously subjected to more heat. It is also a good idea to loosen them a bit in extreme forced induction applications.

Valve Seats

If your application dictates valves larger than 2.02/1.57–inches, larger valve seats must be installed. If you go to a 2.10/1.60–inches combination or larger, the valve seats and guides will also need to be relocated (i.e.: moved away from each other) to allow the valve heads radial clearance. This is a complicated procedure to be performed by a competent machine shop with experience in such operations.

Seat Width—The width of the valve seats is an important consideration because the seats are the sole means by which the valves are cooled. A wide valve seat transfers more heat while the valve is closed, prolonging valve life, especially on the exhaust valve. A narrower seat increases airflow but transfers heat less efficiently. The trick then is to find the right compromise for your engine. The best choice for a street/strip or road race engine is .040 intake seat and .070 exhaust seat.

Once the new seats have been installed, consistent valve heights of 1.800-inches must be established for all of the chambers. This will reduce chamber volume variation and establish a uniform valve stem height. This uniformity requires fewer valve spring shims during final assembly when setting spring height. If adjustments are needed, it is preferable to adjust valve lengths or cut the spring seats. Recessing the seats disrupt airflow.

Valve Job

After seat height is set, the seats need to be prepared to form a seal with the valves. This process is commonly known as the valve job. A

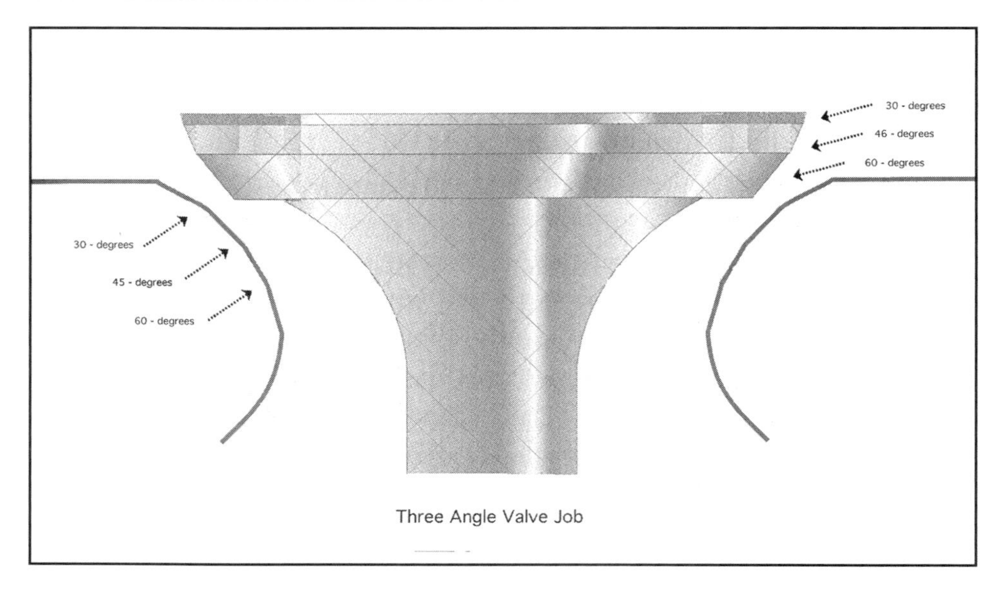

Viton valve seals are the seals of choice for durable, leak-free operation.

three-angle valve job has become the performance standard. Unlike most previous OEM offerings, the LS1 heads are blessed with a three-angle valve job from the factory. By incorporating three angles into the valve and seat, a smoother transition is achieved and airflow improved. Of course, a valve job should be done anytime the valves or seats are changed or modified. Accept nothing less than a premium three-angle valve job for your heads. The factory valve seat angles of 30°, 45° and 60° are ideal.

Valve Seals

Viton has better wear resistance than most other seal materials, making it a good choice for applications where long-term durability is a must. Viton has good flexibility, which means it can handle some run out between the valve stem and guide.

Head Gaskets

The LS1's cylinder head gasket consists of two layers of graphite sandwiching a steel core, and utilizes stainless steel PTFE coated flanges and lacing. The head bolts are a non-reusable torque-to-yield design. The bolts extend 88mm into the main web of the block, giving them very long threads capable of withstanding high load. Fasteners exert the most clamping force when stretched slightly, and the long bolts provide the necessary material to allow this. By anchoring the fasteners deep into the block's backbone, clamping pressure is exerted at the bottom of the sleeves. This tends to draw the deck area immediately surrounding the bores tight against the deck of the cylinder head. For stock bore diameter engines, the GM factory head gasket is hard to beat. But if you are building a big bore, check into top quality

This shot of our 418 super stroker should give you an idea why the custom gaskets are necessary on big-bore engines. Here is a look at a stock head gasket placed on the deck of the assembled engine. As you can see, the piston bore is approximately .050-inches larger than the gasket bore. Copper head gaskets are available, but due to seepage issues, they are generally reserved for all-out racecars.

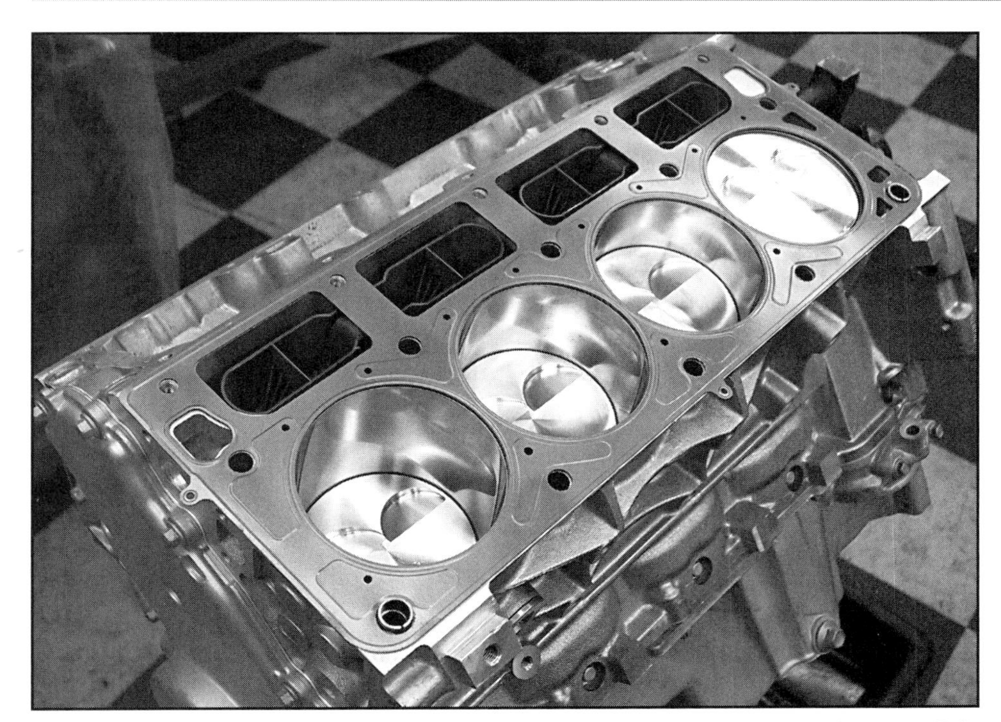

Back to our engine build. Note the Cometic composite head gasket, already in place. This gasket has a bore size of 4.160-inches, which not only works well for this 4.080 bore engine, but is more than adequate to cover the largest of big-bore creations. When compressed, the head gasket is .055-inches thick, which brings the final compression ratio to a very pump-gas friendly 10.9:1. Altering the head gasket thickness is an economical way to change the compression ratio of the engine.

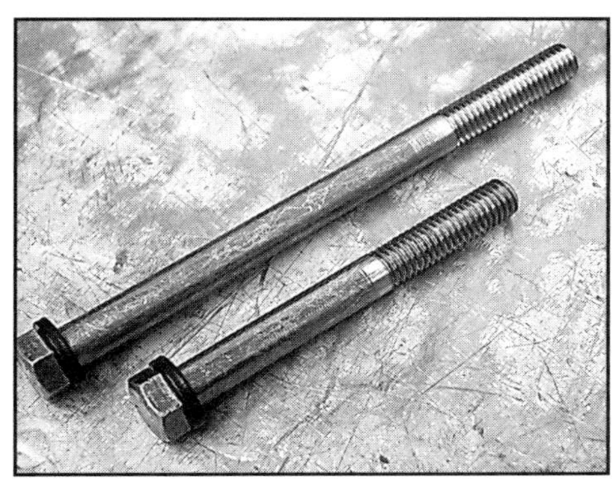

Aftermarket head bolts are a wise investment for an engine destined for use with a power adder.

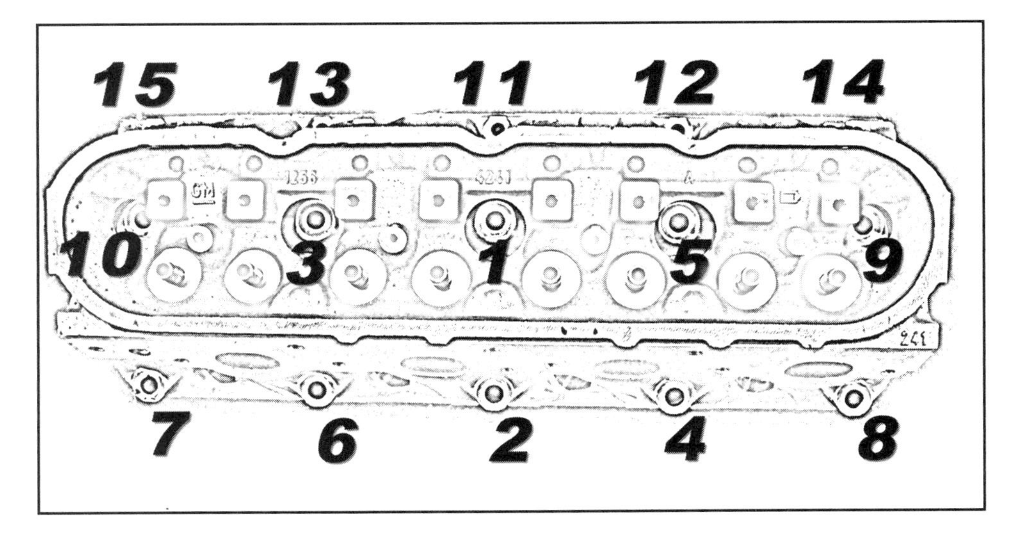

Regardless of the fasteners you select, you will need to follow the torque sequence diagram.

gaskets like those from Cometic.

Head Bolts

A considerable obstacle in assembling a nuke-proof LS1 is the marginal clamping power of the 4 torque-to-yield head bolts per cylinder. More fasteners would allow more even clamping of the cylinder head and lessen the chances of a cylinder head lifting and allowing compression and coolant to slip past the gasket.

Though the stock torque-to-yield bolts work fine in milder naturally aspirated combinations, they're not up to the task of keeping the heads clamped in place when it comes to an extreme build. If your plans include the use of a power adder such as a supercharger or nitrous oxide, upgrading to aftermarket fasteners is a wise investment. Aside from their obvious strength advantage, another benefit of the aftermarket head bolts is that they eliminate the flaky torquing sequence of the factory torque-to-yield hardware. Aftermarket companies such as Automotive Racing Products (ARP) offer head bolts and head studs for the LS1.

Head Studs

Studs provide the same benefits as main studs, namely, they allow greater, more even clamping force. They do, however, limit the ability to remove the heads with the engine in the car.

There are few things that set an enthusiast's heart racing as the sound of a big cam in a high performance engine. The right camshaft not only makes all the right noises, but also is crucial in maximizing the potential of any engine. When the Gen-III engine was unveiled, its conventional, single camshaft, pushrod configuration surprised many people. In a time when nearly all high-performance engines had become overhead cam designs, the LS1's pushrods were something of a throwback. Several factors drove GM to Gen-III's pushrod design, not the least of which was money. Pushrod engines are less expensive to build because they require fewer components and utilize a simpler cylinder head design.

Stock Camshaft Design

The Gen-III camshaft is ground from a 5150 steel billet with induction-hardened lobes and a gun-drilled 17mm centerline bore for weight reduction. The bearing journals, base circle, and body diameters have been increased over previous designs for additional stiffness and easier manufacturing. The camshaft lobe spacing and timing were also altered to accommodate the planar valvetrain geometry and revised firing order.

The use of a larger base circle with less lobe lift reduces valvetrain loading by creating a lower acceleration rate. These lobes employ slower closing ramps that lessen the impact of the valve contacting the seat when closing. This allowed the use of lighter valvetrain parts and lower tension valve springs while improving component service life and reducing valvetrain noise.

The camshaft is supported by five bearings, which are pressed into the cylinder case. In order for the PCM to recognize the position of the engine in the firing order, the camshaft has a machined reluctor ring incorporated between the fourth and fifth bearing journals and a corresponding camshaft sensor located in the block.

Timing Chain

A traditional sprocket and chain assembly drives the camshaft, with both the crank and cam sprockets manufactured from powdered metal. The crankshaft sprocket is splined and

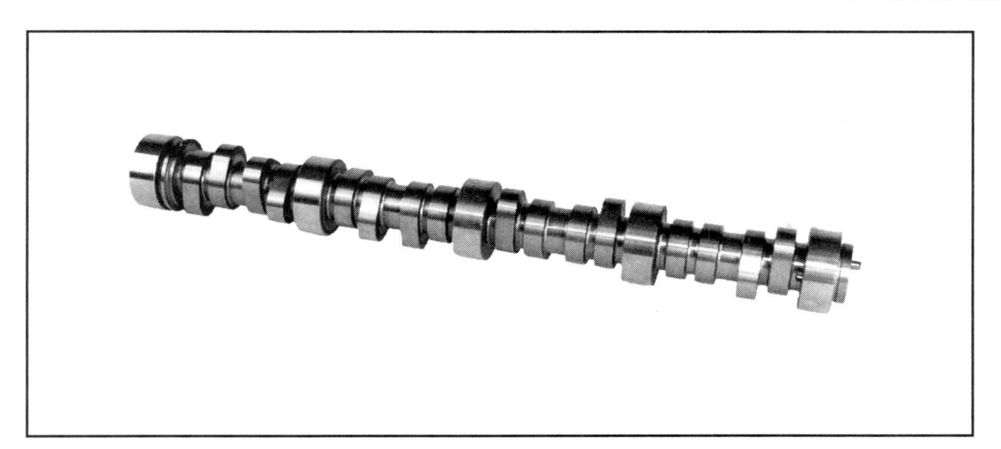

All Gen-III engines employ hydraulic roller camshafts.

Gen-III camshafts have a reluctor ring installed for the camshaft sensor.

Though the stock timing chain has proven durable, even beyond 6000 rpm, upgrading to a double roller like this one from Rollmaster is an excellent idea. Broken timing chains usually lead to mangled valves and all sorts of other mayhem.

The cam sensor is used to track the engine's position in the firing order and located at the rear of the block.

drives the oil pump directly. The chain is a .375-inch (9.52mm) pitch full roller that uses taut strands, so the system does not require a tensioner. A chain guide is utilized to ensure robust performance at high engine speeds. A retaining plate mounted to the front of the engine block prevents fore-aft camshaft walk.

Camshaft Terminology

Far and away the most complex component in an engine is the camshaft. It is the cam that determines when the valves open, how far, and for how long. There are a mind-numbing number of terms and specifications thrown around when discussing camshafts. Before you can make an intelligent decision regarding camshaft selection, you need to understand the basic terminology.

Lift

The camshaft's rotational motion is converted to linear motion by roller lifters following the lobes on the camshaft circumference. Pushrods are seated in the top of the lifters that actuate the rocker arms, which in turn open and close the valves. We all know that opening of the intake valve is what allows the air/fuel mixture into the combustion chamber during the intake cycle, and the opening of the exhaust valve releases the residue of the combustion process during the exhaust cycle. Lift designates the maximum distance the valve is lifted from its valve seat and is measured in thousandths of an inch. Stock Gen-III cams have .475- to .525-inches of lift, while aftermarket cams usually start at .525-inches and go up.

Limiters—It is important to understand that maximum lift is dictated by camshaft duration. In other words, there is a maximum rate of lift that can be generated per degree of cam duration. As duration increases, the time available to open the valve does too. Maximum lobe lift is also limited by the diameter of the camshaft main journals. The height of the lobe cannot be greater than the camshaft journal diameter or the cam won't clear the cam bearings during installation. If lift beyond this threshold is needed, a smaller base circle camshaft can be used.

Duration

Duration is the length of time, measured in degrees of crankshaft rotation, that the valve is open. Though duration is usually measured

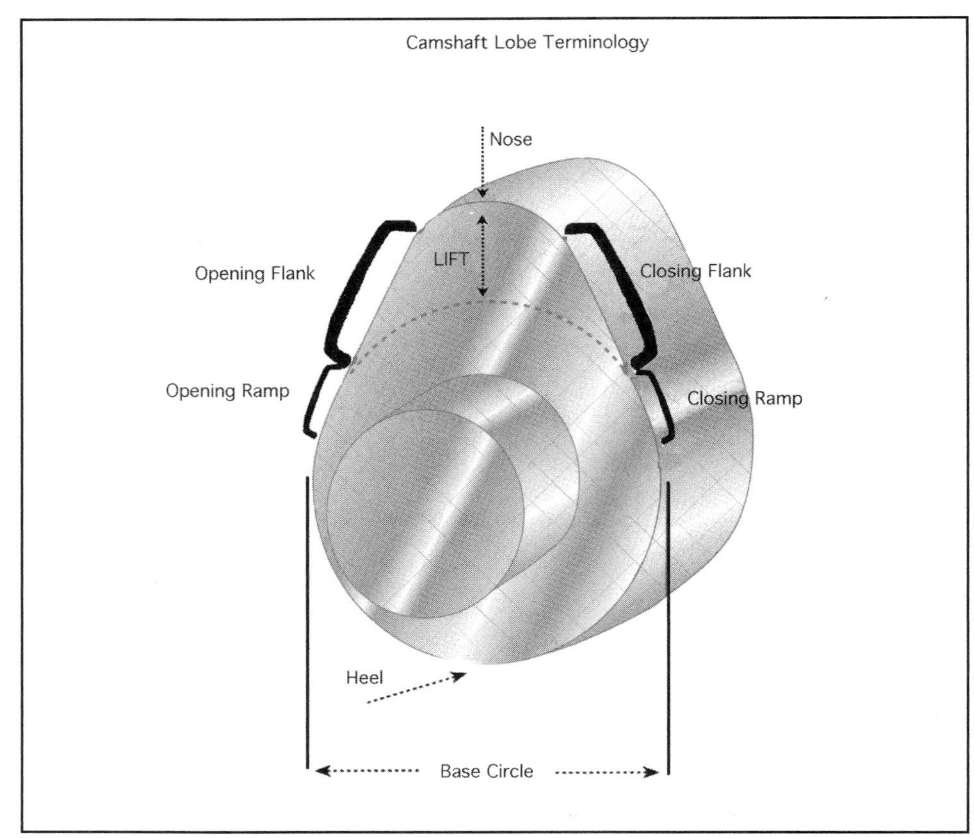

at .050-inches of valve lift, another term you will see is advertised duration. Advertised duration refers to an ambiguous point where tappet movement begins. There is no standard checking point for advertised duration specs, though some cam companies use 0.006-inches of lift. The standardized 0.050-inch of tappet lift makes camshaft comparisons much simpler.

Increased duration means the valve opens sooner and closes later. The longer the valve is held open, particularly the intake valve, the more top-end power the engine will produce because there is more time to fill the cylinder at high rpm. This is why a longer-duration cam will make more peak horsepower than a shorter-duration cam. The downside is that longer duration cams tend to soften low- and mid-range power. This is because the later-closing intake valve allows air and fuel in the cylinder to escape back into the intake manifold at low engine speeds. This condition is known as reversion and occurs because the pressure in the cylinder is greater than the pressure in the intake manifold.

The faster you open and close the valves, the longer you can keep the valve open near max lift, which in turn means more air flow into the cylinder. It has been said, in jest, that the ideal camshaft would have a square lobe that slams the valves open and closed at maximum velocity. Of course, this isn't practical in the real world, as slamming the valve closed at maximum velocity would make for ridiculously short valvetrain component life. A valve opens and closes more than 100 times per second at engine speeds over 6000 rpm. And each time the valve closes, there is a collision of metal parts as it strikes the valve seat. By slowing down the rate of closing as the valve approaches the seat, the strain on cylinder head and valvetrain components caused by this impact is lessened.

Stock Gen-III cams generally have between 200° and 210° of intake lobe duration at 0.050-inch valve lift, depending upon the application. Smaller performance camshafts start at around 210°. For a street and strip car 220° to 230° works well. Durations of 230° or more are best left for the aggressive stroker engines or racecars.

Lobe Separation Angle

Lobe Separation Angle (LSA) is the angular displacement between the centerlines of the intake lobe and its companion exhaust lobe. LSA is sometimes referred to as lobe center angle. As lobe separation is increased, overlap is decreased (see below). Because it pertains to lobe-to-lobe displacement, this is one of the few instances when a cam dimension is given in cam degrees rather than crank degrees.

There is an LSA that will produce maximum horsepower for any given engine and camshaft profile. A wide lobe separation angle in the 116° range will provide a smooth idle, a broad torque band, and reduce the engine's sensitivity to a less than perfect combination. Tightening the lobe separation angle to 106° would ultimately provide more power, but at the expense of a narrow power band and a choppy idle. This is why cams with very tight LSA are better suited to race-type applications. The LS1 uses a very wide lobe separation angle (as much as 119.5° in some applications) to produce a silky smooth idle. Lobe separation angle in performance cams usually falls between 102° and 116°, with 110° and 112° being very good all-around performers in the LS1.

When choosing an LSA for anything less than a full-blown racecar, it is best to broaden the power curve by using a conservative LSA. LSAs of 110° to 114° are usually best for engines that will see street duty as well as the occasional foray at the track. A wider lobe separation angle improves idle quality by creating more intake manifold vacuum and allowing less reversion. Stick close to this sweet spot, and you'll have a responsive engine that makes the car a blast to drive.

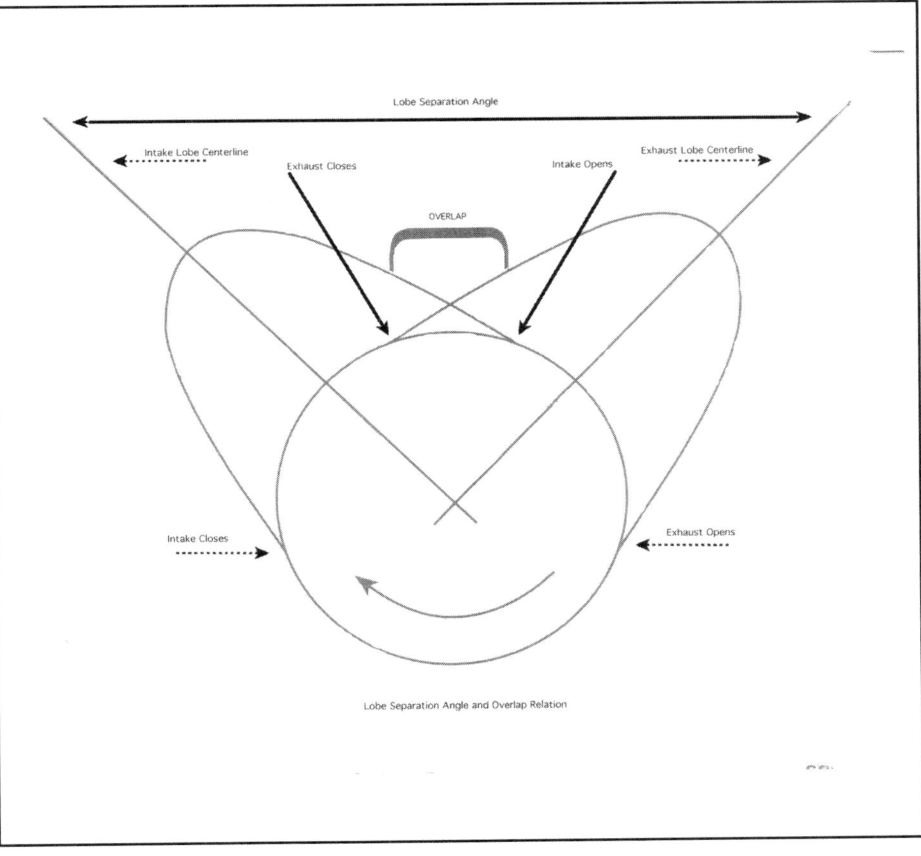

Lobe Separation Angle and Overlap Relation

Intake Centerline

Intake Centerline refers to the position at which the cam is installed relative to the crankshaft. It tells us the angle, in crankshaft degrees, at which maximum intake valve lift occurs relative to Top Dead Center (TDC). For example, an intake centerline of 108° tells us that the intake valve reaches maximum lift at 108° after TDC. This information is useful in verifying the position and dimensions of the camshaft (see Camshaft Installation, p. 84).

Overlap

Many people credit a cam's lumpy idle to increased intake duration, but overlap plays a bigger role. Valve overlap is measured in crankshaft degrees. It occurs as the piston approaches and leaves TDC extending from the exhaust stroke of one cycle to the intake stroke of the next. Lobe Separation Angle (LSA) directly affects overlap.

Increasing the LSA decreases the amount of overlap. However, if the intake-lobe centerline remains the same, then spreading the lobe separation angle has the effect of leaving the intake lobe in place and advancing the exhaust-lobe centerline. Conversely, decreasing LSA increases overlap. This improves the volumetric efficiency through the engine's midrange rpm by allowing better exhaust scavenging. Increasing overlap with a short-duration cam is a way to increase torque in the mid-range, but sacrifices torque at lower rpm.

Valve overlap is also affected by altering camshaft duration. As duration is increased, valve overlap will increase, providing the LSA is not modified to compensate for the increased duration. A longer-duration camshaft with the same lobe separation angle increases the amount of overlap by a substantial amount. This is why some longer-duration cams use wider lobe separation angles.

Valve Event Timing

The points at which the valves open and close relative to the position of the crankshaft are obviously very important in optimizing performance. Cam timing is described in the valve event data provided on the cam card. Duration, valve overlap, event timing, and lobe center angles are all interrelated, and it is not possible to tune each of these characteristics independently for single cam engines. Because of the tradeoff nature of cam timing, if valve duration and overlap are increased past a certain point, all additional top-end power will be produced at the expense of low-end performance. Where the threshold lies depends on engine displacement, compression ratio and several other factors.

Choosing the right cam for your engine should involve no guesswork.

There are four opening and closing points: intake opening (IO), intake closing (IC), exhaust opening (EO), and exhaust closing (EC). All four points should be specified on the cam card. The most important of these is the intake closing point because it determines the point at which the rising piston starts to build cylinder pressure. High cylinder pressure results in improved torque production.

Single-Pattern or Dual-Pattern?

Single pattern camshafts have identical duration and lift specs for the intake and exhaust lobes. Dual- or split- pattern cams have more exhaust duration and lift than the intake side or vice-versa. Dual-pattern cams are a method to address the limited breathing capacity of a cylinder head's exhaust port. Increasing the exhaust valve lift and duration helps evacuate exhaust residue from the cylinder at high rpm when there is less time to achieve this. Whether your engine would benefit from a dual pattern cam depends on your cylinder heads' flow capacities and E/I ratio (see Chapter 5). Engines with balanced cylinder head flow characteristics usually perform best with a single pattern cam.

Solid rollers require the use of adjustable rockers to adjust valve lash. These shaft mount T&D rockers are available in many ratios.

Performance Camshafts

As with virtually every other performance modification, there are many variables that require consideration in the cam selection process, including but not limited to: engine displacement, cylinder head flow capabilities, forced induction, torque converter selection (if the car is an automatic), rear gears, and weight of the car. Always remember that a cam is a give-and-take component. Aggressive cams make more power but move the power peak higher in the rpm range. There is a point where all top-end gains are made at the expense of low-end output, resulting in a loss of driveability and fuel economy. And in order for a larger cam to make more power, the cylinder heads need to support the additional airflow. It doesn't matter how long the valves are open if the heads are incapable of providing sufficient air.

The development of the modern hydraulic roller cam allowed cam designers to incorporate the advantages of the more aggressive flat tappet mechanical cam into a very driveable, low-maintenance package. There is no denying that modern hydraulic rollers are capable of producing some amazing power numbers.

For an engine that will see a lot of street miles, it is tough to argue against a hydraulic roller cam. A

How to Set Lash on a Solid Roller

If you opt for a solid cam, you will need to set the valve lash occasionally. Lash is the distance between the valve tip and the tip of the rocker arm. It's the "slop" in the valvetrain that must be taken up before the valve begins to move. This periodic maintenance is a good idea prior to every trip to the track. Once the valvetrain is past its initial break-in, the lash values should not change by more that a couple thousandths of an inch. Dramatic changes could be indicative of valvetrain issues: The rocker arm may not have been tightened enough during installation or valvetrain damage may occur.

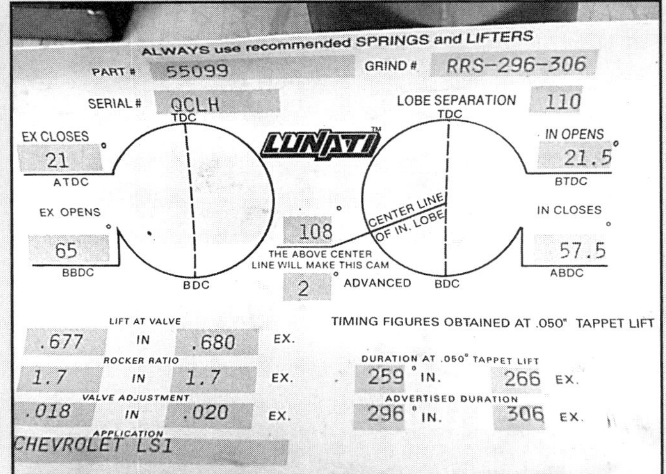

ALWAYS use recommended SPRINGS and LIFTERS
PART # 55099 GRIND # RRS-296-306
SERIAL # QCLH LOBE SEPARATION 110
LUNATI
EX CLOSES 21° ATDC
EX OPENS 65° BBDC
IN OPENS 21.5° BTDC
IN CLOSES 57.5° ABDC
TDC BDC
CENTER LINE OF IN. LOBE 108°
THE ABOVE CENTER LINE WILL MAKE THIS CAM 2° ADVANCED
LIFT AT VALVE .677 IN .680 EX.
ROCKER RATIO 1.7 IN 1.7 EX.
VALVE ADJUSTMENT .018 IN .020 EX.
APPLICATION CHEVROLET LS1
TIMING FIGURES OBTAINED AT .050" TAPPET LIFT
DURATION AT .050" TAPPET LIFT 259° IN. 266 EX.
ADVERTISED DURATION 296° IN. 306 EX.

Valve lash is usually set around .024 inches but refer to the cam card for exact specifications. There is some play in this value; increasing the lash will make it act like a smaller cam and will add a slight gain in low-end torque production. Tightening the lash will slightly increase high-end power. However, in most applications, the slight increases are not worth the extra risk of parts breakage.

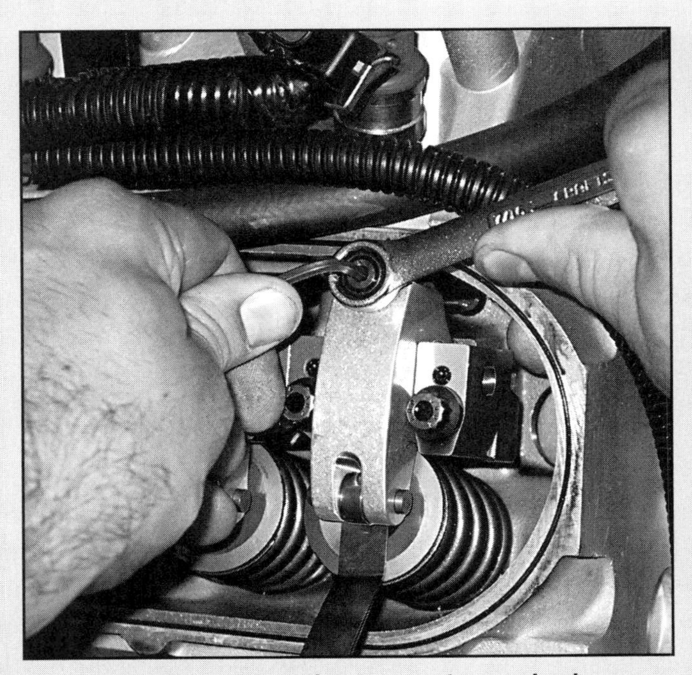

With the engine at operating temperature and valve cover removed, a feeler gauge is inserted between the valve tip and the roller on the rocker arm. The lash is usually set with an Allen key and 12 point wrench. The intake is set when the exhaust is just beginning to open and the exhaust is set when the intake is beginning to close.

The engine will need to be bumped over during the process, which turns the valvetrain in order to check all the valves. Consider the addition of an ignition bump switch, which does the bumping and makes running the valves a one-person chore.

Always use a generous amount of engine oil when installing a camshaft, and exercise caution so as not to nick the cam bearings.

This engine uses a Cloyes Hex-A-Just timing chain. This timing set allows the cam to be advanced or retarded without the need for offset keyways.

strong hydraulic grind in a super stroker is easily capable of making in excess of 500 horsepower. This, combined with the zero-maintenance aspect of hydraulic lifters, make this a no-brainer for a street engine. But when maximum power is required, a solid roller is the way to go.

Mechanical Roller Cams

Mechanical camshafts (more commonly referred to as solid cams) are not a new idea. Many of the hottest engines from the musclecar era came factory-equipped with solid cams. These cams employed more aggressive lobe profiles and required the use of meatier valve springs to eliminate valve float at high rpm. They also required an adjustable rocker arm for valve lash adjustments.

The advantages of a solid roller cam are numerous. Because it incorporates very fast ramp speeds, a solid roller will snap the valves open and closed much quicker than a hydraulic cam. High-rpm valvetrain stability—over 6000 rpm—is another benefit. This is where hydraulic lifters often begin to pump up, costing considerable power.

The disadvantages of a solid roller cam are comparatively few. Many assume that lashing the valves is a difficult and time-consuming proposition. On the contrary, adjusting the valves is really quite easy. If you can change your own spark plugs, you can handle lashing the valves. It is unavoidable that the valvetrain makes a bit of clatter. Because this noise is of a frequency similar to that of cylinder pre-ignition, the knock sensors may need to be desensitized in the PCM program. This is not a problem on a naturally aspirated engine, but due care needs to be exercised to ensure the engine has sufficient octane at all times. It goes without saying that this becomes even more critical in a power-adder situation.

An interesting aspect of mechanical cams is that they behave similarly to a smaller hydraulic. This is because the fast ramps decrease the amount of time the valve is off the seat thereby bleeding off less cylinder pressure. Solid cams often increase low-rpm torque, which translates directly to improved drivability compared to a similarly sized hydraulic cam.

Camshaft Installation

In theory, the installation of a camshaft is a very straightforward task: stab the cam in the block, line up the dots on the timing set, and button it up. The proper way to finish the job is by degreeing the cam. Because a camshaft cannot be inspected prior to installation with a caliper or micrometer, a cam is "degreed-in" to check the accuracy of the grind. In the process of degreeing a camshaft, you're verifying that valve opening and closing events match the specifications found on the cam card. Valve opening and closing events are influenced not only by the accuracy with which a cam was manufactured, but the accuracy of the timing chain set and deflection in other valvetrain components. Degreeing a camshaft is not a difficult procedure, but it does require some attention to detail and a few specialized tools.

Locate TDC

The first step in degreeing the

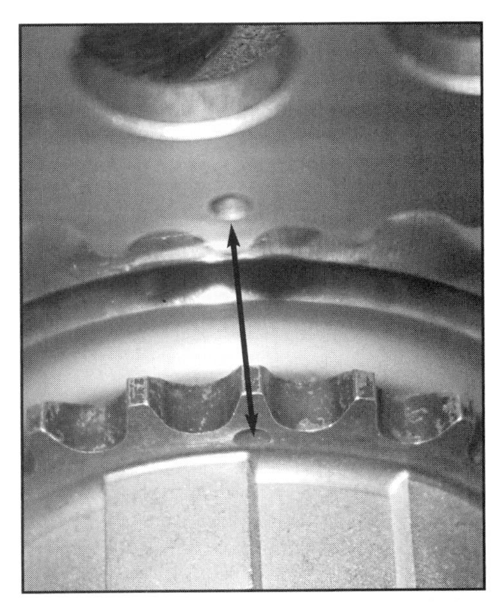

With the camshaft in place, install the timing chain, being sure to align the dots on the timing chain sprockets. The cam sprocket dot should be at 6 o'clock and the crank sprocket dot at 12 o'clock.

A degree wheel is then bolted to the front of the crank. Put the wheel's 0° or TDC mark at 12 o'clock and fasten it with the crank bolt.

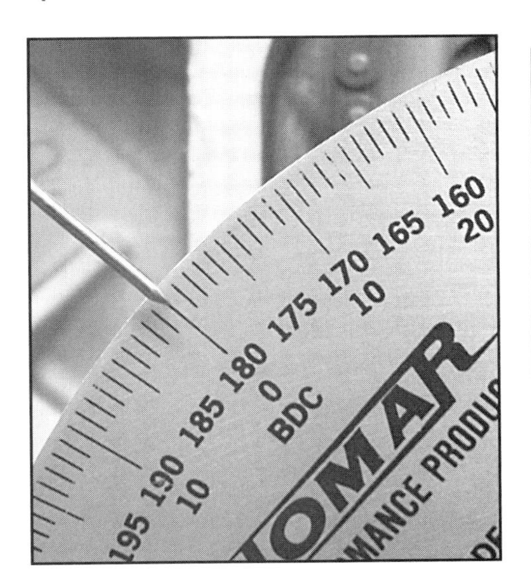

A length of wire coat hanger makes a fine pointer. After cutting an appropriate length of wire, sharpen one end on a grinding wheel (or use a file), as this will provide an accurate pointer. Choose a timing cover bolt hole to which the pointer will be installed. Bend the wire to a convenient position on the degree wheel.

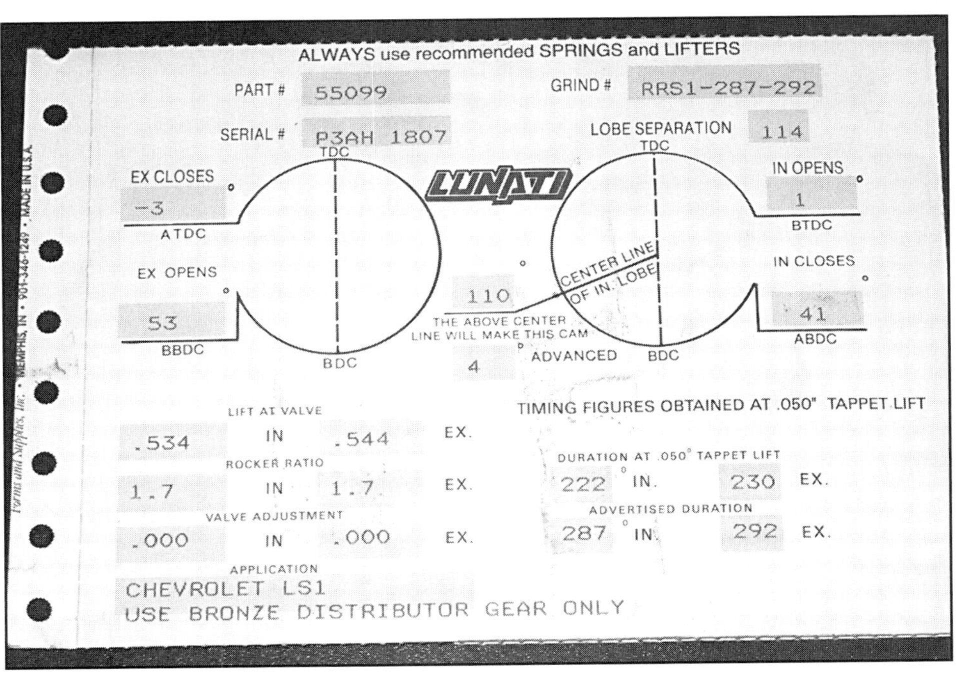

ALWAYS use recommended SPRINGS and LIFTERS

PART # 55099 GRIND # RRS1-287-292

SERIAL # P3AH 1807 LOBE SEPARATION 114

LUNATI

EX CLOSES -3 ° ATDC

EX OPENS 53 ° BBDC

IN OPENS 1 ° BTDC

IN CLOSES 41 ° ABDC

CENTER LINE OF IN. LOBE 110 °

THE ABOVE CENTER LINE WILL MAKE THIS CAM 4 ° ADVANCED

TIMING FIGURES OBTAINED AT .050" TAPPET LIFT

	IN		EX
LIFT AT VALVE	.534	.544	
ROCKER RATIO	1.7	1.7	
VALVE ADJUSTMENT	.000	.000	
DURATION AT .050" TAPPET LIFT	222°	230	
ADVERTISED DURATION	287°	292	

APPLICATION
CHEVROLET LS1
USE BRONZE DISTRIBUTOR GEAR ONLY

Cam manufacturers include a cam card with their products. On it is all of the data you will need to degree the camshaft. This card will be used in this example.

camshaft is to accurately locate the engine's top dead center (TDC). If the heads are on the block, use a piston stop through the spark plug hole on the # 1 cylinder to locate TDC. If the heads are off, a top of the block piston stop will be required. Once the piston stop is in place, it's time to find TDC.

Remove the piston stop and turn the crank until the degree wheel arrives at TDC on the pointer. Be careful not to disturb the pointer.

If, as in the case of our Lunati example, the lift numbers on the cam card refer to lift at valve, you will need to multiply the lobe lift number by rocker arm ratio.

Find TDC—Rotate the crankshaft slowly clockwise until the piston stops against the bolt head. Next, reposition the wire pointer to 0° or TDC on the degree wheel. Then, rotate the engine counterclockwise (opposite direction) until the piston again reaches the stop. Note the reading on the degree wheel and divide the figure by two. For example, if the degree wheel shows 32°, half of that figure is 16°. Move the pointer to the number just computed (16°) on the wheel. Now, if the calculation is correct, by turning the crank the opposite direction it should stop on the same number on the other side of TDC (i.e., 16°). If the two numbers (e.g., 16°) are the same, TDC has been located correctly.

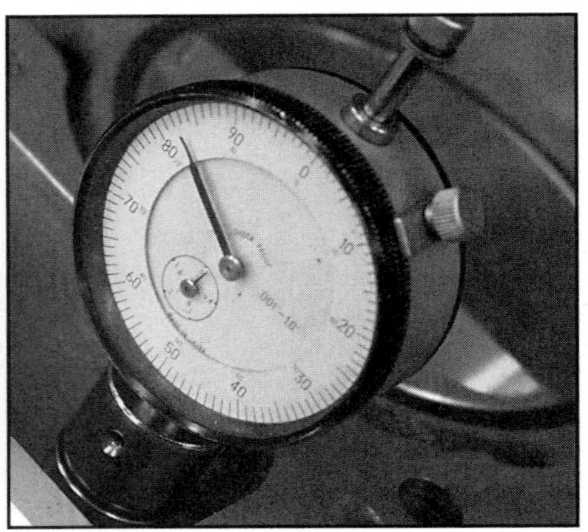

Verify Lift—Before checking cam timing, verify cam lift. Install a lifter on the #1 intake lobe and set up the dial indicator so that indicator shaft sits on the body of the lifter (not in the cup). It is good practice to spin the crank several times to be sure that the dial indicator is giving repeatable readings. Once the measuring apparatus is properly aligned, rotate the crank clockwise until the lifter is on the heel (base circle) of the cam lobe. This is the lowest point, so the dial indicator should be set to zero. Begin rotating the crank until reaching maximum lift. Note the lift number and compare it to the cam card. If measured lift doesn't match up with the specs on the cam card, the dial indicator may not be set up properly. Verify that the dial indicator is aligned perfectly with the lifter and repeat this step.

Verify Duration—Begin with the lifter on the cam's base circle with the dial indicator reset to zero. Rotate the crank in its normal, clockwise direction. As the needle passes .040-inches of lift, slow the rotation so that the needle approaches the .050-inches indication slowly. Once .050-inches is attained, note the reading on the degree wheel, it should correspond to the "INTAKE OPENS" specification on the cam card (1° Before TDC). It is imperative that you arrive at .050-inches by spinning the crankshaft in its normal direction. Going past the desired point and then backing up will result in incorrect opening and closing point readings due to lack of preload on the timing chain.

Example: .544 / 1.7 = .320 inches of cam lift

With the opening point noted, continue to rotate the crankshaft. When the piston begins its descent, slow again to sneak up on .050-inches lift. When it is reached, again note the reading on the degree wheel. It should agree with "INTAKE CLOSES" on the cam card (41° After Bottom DC). It is good practice to repeat the entire procedure to verify the results. Once satisfied, perform the procedure on the cylinder's exhaust lobe.

Intake Centerline

An alternative method of checking cam phasing is by the Intake Centerline Method. Intake centerline is the point at which the intake lobe reaches maximum lift. Typically, high-performance cams reach intake maximum lift at 100° to 109° After TDC. The problem with using the centerline method is that the actual opening and closing points are not checked, so it is unlikely to detect machining inaccuracies in the cam.

Many engine builders use intake centerline as a reference point when altering camshaft timing to shift an engine's power curve. It is best to calculate intake centerline by dividing intake duration by 2, then subtracting the intake opening number. See the example below, based on our example Lunati cam:

Example: (222° / 2) – 1° = 110°

Based on this number, the installer can decide whether they wish to advance or retard the cam timing using an adjustable timing set.

GM Camshaft Specifications, by Application

	Part Number	Valve Lift (in., e/i)	Duration (e/i)	Lobe Separation Angle (degrees)	Timing
1997 LS1 Y-car	12554710	.479/.472	207/199	117	113/121
1998 LS1 Y-car	12554710/12560964*	.479/.472	207/199	117	113/121
1999 LS1 Y-car	12560964	.479/.472	207/199	117	113/121
2000 LS1 Y-car	12560968	.500/.500	209/198	115.5	112/119
2001 LS1 Y-car	12561721	.479/.467	207/196	116	117/115
2002 LS1 Y-car	12561721	.479/.467	207/196	116	117/115
2003 LS1 Y-car	12561721	.479/.467	207/196	116	117/115
2001 LS6 Y-car	12560950	.525/.525	211/204	116	114/118
2002 LS6 Y-car	12565308	.551/.555	218/204	117.5	115/120
2003 LS6 Y-car	12565308	.551/.555	218/204	117.5	115/120
1998 LS1 F-car	12557812/12560965*	.500/.500	209/198	119.5	117/122
1999 LS1 F-car	12560965	.500/.500	209/198	119.5	117/122
2000 LS1 F-car	12560965	.500/.500	209/198	119.5	117/122
2001 LS1 F-car	12561721	.479/.467	207/196	116	117/115
2002 LS1 F-car	12561721	.479/.467	207/196	116	117/115
2001 LS1 V-car	12560965	.500/.500	209/198	119.5	117/122
2002 LS1 V-car	12560965	.500/.500	209/198	119.5	117/122
2003 LS1 V-car	12560965	.500/.500	209/198	119.5	117/122
1999 LR4	12560966/12560967*	.466/.457	190/191	115.5/114	118/113 112/116
2000 LR4	12560967	.466/.457	190/191	114	112/116
2001 LR4	12560967	.466/.457	190/191	114	112/116
2002 LR4	12560967	.466/.457	190/191	114	112/116
2003 LR4	12560967	.466/.457	190/191	114	112/116
1999 LM7	12560966/12560967*	.466/.457	190/191	115.5/114	118/113 112/116
2000 LM7	12560967	.466/.457	190/191	114	112/116
2001 LM7	12560967	.466/.457	190/191	114	112/116
2002 LM7	12560967	.466/.457	190/191	114	112/116
2003 LM7	12560967	.466/.457	190/191	114	112/116
2003 LM4	12560967	.466/.457	190/191	114	112/116
1999 LQ4	12560967	.466/.457	190/191	114	112/116
2000 LQ4	12560967	.466/.457	190/191	114	112/116
2001 LQ4	12561721	.479/.467	207/196	116	117/115
2002 LQ4	12561721	.479/.467	207/196	116	117/115
2003 LQ4	12561721	.479/.467	207/196	116	117/115
2002 LQ9	12561721	.479/.467	207/196	116	117/115
2003 LQ9	12561721	.479/.467	207/196	116	117/115

*Change in GM Part number only, did not affect cam specs.

7 Valvetrain

The mass and stiffness of the LS1 valvetrain was optimized by GM Powertrain to achieve engine performance, noise, and fuel economy and emissions goals. The valvetrain geometry was designed to be virtually planar in both the longitudinal and axial directions, resulting in pushrod angles of less than one degree relative to the lifter bore centerline. To achieve this ideal geometry, the camshaft centerline was raised in the block and new lobe spacing was used. These efforts allow the stock LS1 valvetrain to perform reliably at engine speeds in excess of 6000 rpm.

Lifters

Lifters are the means by which the rotary motion of the camshaft is converted into linear motion. The LS1 utilizes hydraulic roller lifters. Roller lifters require a billet-steel cam that is extremely hard compared to a cast-iron flat tappet cam. Because roller lifters roll over the lobe, they are robust under heavy load and at high engine speed. Roller lifters allow the use of lobes with high lift and fast ramps; cams ground with these characteristics make big power!

Hydraulic lifters automatically adjust valvetrain lash by supporting a pushrod plunger on a small chamber of pressurized engine oil, allowing them to compensate for heat expansion and component wear. The chamber uses a controlled bleed orifice that allows the plunger to move up and down slightly, preloading the system, maintaining zero lash, and making them virtually maintenance free.

Lifter Pump-Up

The only real drawback to hydraulic lifters is their tendency to pump up at high rpm. When this happens, the lifter prevents the valves from closing completely, bleeding off cylinder pressure and greatly diminishing power production.

Solid Lifters

Solid, or mechanical, lifters appear similar to their hydraulic cousins and perform the same function, but do so in a simpler manner. As their name implies, solid lifters have solid bodies that forego the cushioning oil chamber to automatically maintain valve lash. Therefore, engines using

Hydraulic lifters (L) are nearly indistinguishable from the solid variety.

Crane Cams offers solid lifters that allow the continued use of the OEM lifter retainers. With the pushrods removed, these retainers will literally "grip" the lifters thereby allowing a camshaft change without removal of the cylinder heads.

solid cams and lifters require manual lash adjustment through the use of adjustable rocker arms. Solid lifters replace the stock lifters with no modifications and allow the continued use of the OEM plastic lifter retainers. The advantages of a solid roller setup include more reliable high-rpm operation and their tolerance of very high valve spring pressures, which in turn allows very aggressive camshaft profiles to be used.

Pushrods

If stronger valve springs and higher rpm are part of your plan, upgraded pushrods should be, too. The centerlines of the lifters, pushrods and the valve stems are within 1° of parallel. This in-line or planar valvetrain geometry reduces friction and allows some parts to be made smaller and lighter. Pushrods are one such item.

Unless the engine has been over-revved, a bent pushrod is almost always indicative of a valvetrain geometry problem. Weak valve springs that allow high-rpm valve float can also result in bent pushrods when the valves crash into the pistons. Pushrods are designed to support a load in a linear fashion, and they don't live long when subjected to side loads. To ensure pushrod longetivity, verify that they are straight, the correct length, and that they don't encounter obstacles during their motion.

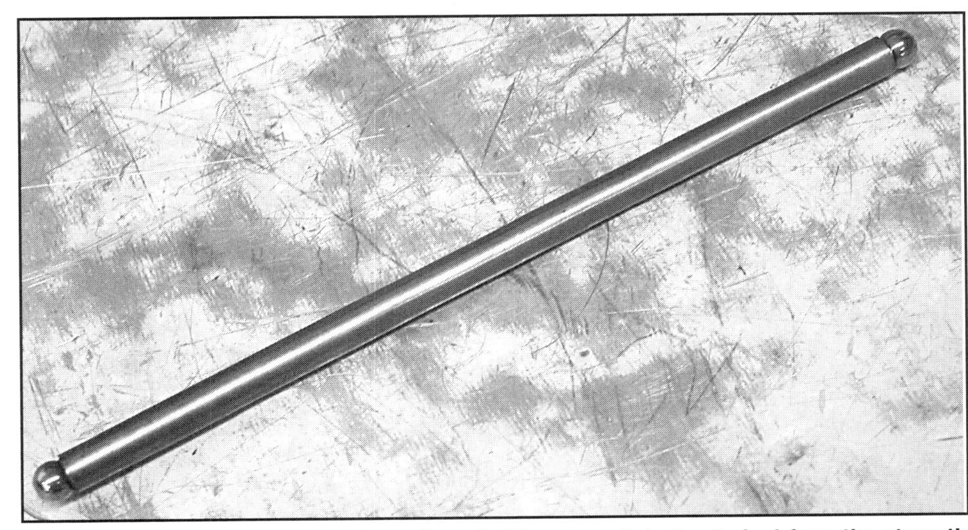

Stock pushrods are okay for stock engines, but leave much to be desired from the strength perspective for high performance builds.

Special length pushrods are necessary only when stock-length pieces do not allow proper valvetrain geometry. Pushrod length will vary significantly from engine to engine. Some of the variables affecting length are whether the heads have been milled, the block decked, the camshaft's specs, lifter type and valve length. Features you will want to look for are diameter of at least 5/16-inch, .080-inch wall

The key to this verification is noting the contact patch of the intake rocker arm as you rotate the cam. You will need to examine it at zero lift, half lift and maximum lift. At zero lift the contact patch should not reach the inside of the valve tip. If it is too close to the inside edge of the valve, the pushrod is too short.

thickness, and 4130 chrome-moly construction.

Rocker Arms

Rocker arms multiply cam lobe lift by a set ratio in order to increase valve lift. The Gen-III rocker arm has an investment-cast steel body that pivots on a needle roller bearing assembly. The inherent versatility of the investment casting process allowed minimization of rocker mass for extremely low rotational inertia. The roller bearing pivot reduces friction and noise. The rocker arm is mounted on a die-cast aluminum rocker arm stand that provides the proper pivot height and alignment with the valve tip. An 8mm bolt is used to locate the rocker arm and fasten it directly to the cylinder head through a clearance hole in the stand.

Rocker Arm Ratio

Rocker ratio is a comparison between how much the valve moves relative to how much the pushrod moves. Rocker ratio is calculated by dividing the distance from the rocker pivot centerline to the pushrod cup

How to Check Valvetrain Geometry

Rocker arm geometry should be verified whenever any valvetrain components are changed. This is easier to check before engine assembly, but can be done after the fact, too. If the engine is not assembled, you will have to put together at least one cylinder. It's best to use lighter-weight checking springs in this step.

At the halfway point, when the valve is opened to about half the maximum, you should expect it to be very close to the center of the tip. At maximum lift, verify that it has not reached the outside edge of the valve stem. If the contact patch moves too close to the outside edge or off the end of the valve stem, the pushrod is too long. This is common if the heads or block were milled for increased compression.

Why Does the LS1 Bend Pushrods?

If there is a single complaint most commonly heard from LS1 owners, it is regarding bent pushrods. While most who have had this problem will admit to blowing a shift with their six-speed, they argue that the rev-limiter should have protected the engine. This is nonsense.

Let's speak hypothetically for a moment: Imagine you are winding the car out in second gear. Here comes your 6000-rpm shift point, you keep your right foot planted and just tickle the clutch pedal while you stuff the shifter right back into first gear. Oops, that's no good. Much mayhem emanates from underhood, and you know you've got problems.

So what happened here? By shoving the transmission back into first gear, you mechanically forced the engine well above 7000 rpm. A rev limiter can only do its job if an engine is being accelerated through use of the throttle. In particular, the LS1's PCM will alternately pull fuel from various cylinders to limit the engine's rpm. It cannot possibly slow down an engine that is being mechanically over-revved through the transmission because of a blown shift.

Because the OEM valve springs are relatively weak, they are unable to keep the lifters in contact with the camshaft lobes under high rpm situations. This leaves the valves hanging open, a condition commonly referred to as valve float. When this occurs, it inevitably allows the pistons to contact open valves that would otherwise be closed.

The result of all this is a fistful of bent or broken pushrods and, if you're particularly unlucky, a couple of bent valves and broken valve springs thrown in just for good measure. Voila: broken engine. If you must blame a component, make it the shifter!

Stock rocker arms are investment-cast steel with a 1.7:1 ratio and feature roller bearing fulcrums.

SLP's 1.85 ratio rocker arms are based on the stock rockers and are built by the same company that builds the stockers for GM. That means they are using the same high quality materials and processes for great durability.

centerline, by the distance from the pivot centerline to the center of the valve cup or roller tip. The geometry of the OEM rockers is Gen-III specific and utilizes a 1.7:1 rocker arm ratio.

Changing rocker ratio is a great way to gain valve lift without changing camshafts or significantly altering the lift duration. Increasing valve lift is a good way of increasing power because it can do so without substantially affecting low-speed performance. For example, a 2001 LS6 cam has an intake lobe lift of .309 inches. Multiply this by the rocker ratio of 1.7:1 to get the valve lift of 0.525 inches. If you change the rocker ratio to 1.85:1, the lift increases to 0.572 inches. Be aware that this method is an approximate, as it does not take into account valvetrain flex. A valve spring upgrade is mandatory when increasing valve lift by any means, including increased rocker arm ratio.

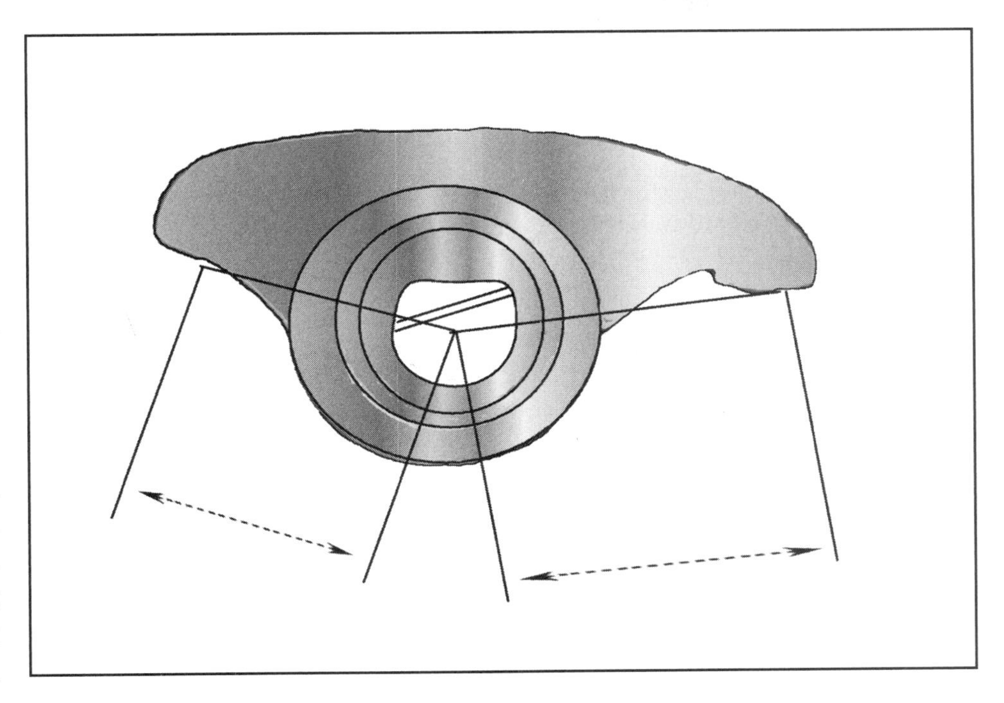

Rocker arm ratio is determined by dividing the distance from the pivot to the pushrod cup by the distance from the pivot point to the valve cup.

Roller Rockers

Another upgrade option is to bolt on a set of roller rockers. Rollers might seem like overkill on a street engine, but they do offer several real-world advantages. Because the operating tip of a roller rocker rolls across the tip of the valve, side loading on the valve stem is dramatically reduced; this increases valve stem and guide life. And since high valve lifts can greatly increase side loads and wear, rollers often make high lift cams more practical for street use.

Installing 1.85:1 Rocker Arms

By increasing the rocker arm ratio, you get the benefits of a higher lift cam without the tuning headaches that often accompany installation of a bigger stick. The increase over the stock rockers' 1.7:1 ratio doesn't sound like much, but a 1.85:1 ratio rocker increases valve lift by nearly 10% and accelerates the valves more quickly during opening and closing. SLP's rocker was designed as a bolt-on replacement for stock. No valve cover or cylinder head modifications are necessary and stock pushrods can be reused. For the results, see the dyno test on page 94.

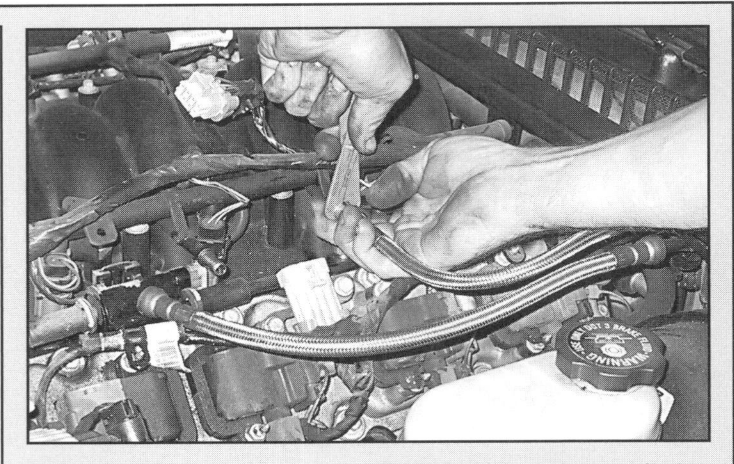

Swapping the rockers, valve springs and retainers can easily be accomplished in an afternoon, provided you have the necessary tools. Once the fuel rail covers are out of the way, the fuel lines should be disconnected and moved aside. The correct way to remove the fuel lines from the rail is with a tool made specifically for the job, as shown here. These tools are reasonably priced and available at nearly any auto parts store.

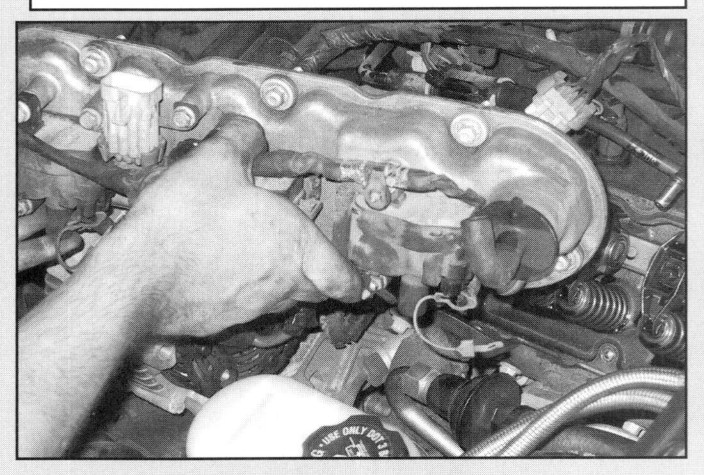

After removing the plug wires and disconnecting the coil pack harness, the valve cover can be removed. It should go without saying that you want to avoid dropping any dirt or debris from the top of the covers into the engine!

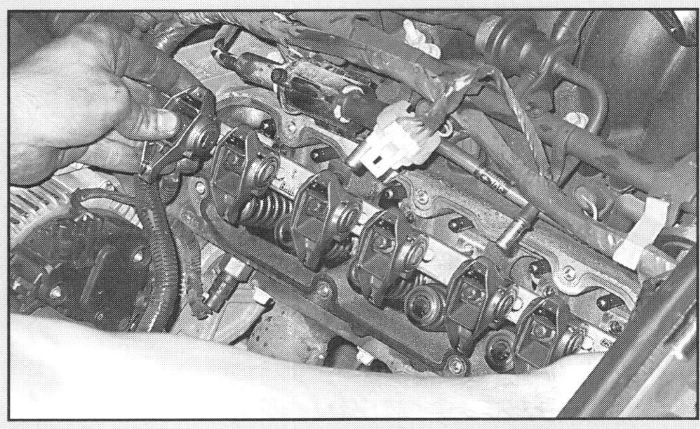

The rockers and rocker stands can be removed as a single unit once the bolts are removed.

Install a spark plug port adapter into the first cylinder spark plug hole and connect an air hose and a source of compressed air. With the cylinder pressurized, the valve spring can be compressed and the locks removed. A pen magnet works best for this. Extreme care should be taken to ensure the locks are not dropped into the head.

The stock honeycomb spring (left) is not up to the demands of increased valve lift. SLP's springs (right) provide 105 – 115 lbs. of seat pressure at an installed height of 1.800-inches, and 310 – 340 lbs. at 1.200-inches. The springs measure 1.250-inches in diameter and are constructed from Cr-Si that allows for lift up to .600-inch. These springs will allow higher RPM due to their increased spring rate and require no head machining to install.

SLP offers their rocker arm packages in two flavors, the difference being the material from which the valve spring retainers are constructed: chro-moly and titanium. Though more expensive, titanium retainers are significantly lighter, which makes the valve springs' job easier, and will allow the engine to rev more freely.

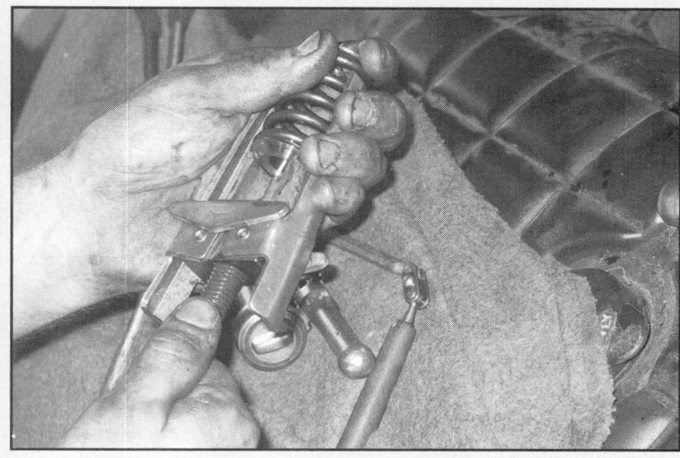

While there are countless types of valve spring compressors available, this style is one of the best for the job on an LS1. SLP recommends the GM compressor, P/N# J 38606. Use whichever you have access to.

Reassembly is a snap. Be sure to torque the rocker arm bolts to 22 ft.-lbs. The increased rocker ratio (over 1.7 stock) increases valve lift from .497-inches to .541-inches. The LS1 heads are definitely good enough to make use of the extra air!

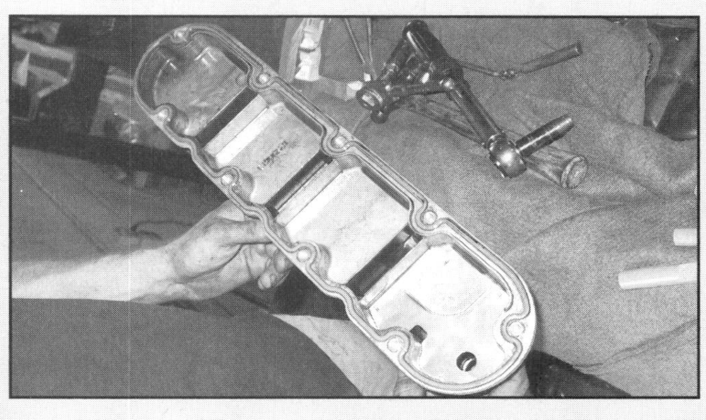

Though the silicone valve cover gaskets are reusable and really durable, take a minute to inspect them before bolting them back in place. Replacements are as close as the local GM parts department.

With everything buttoned up, it was time to make some noise on the dyno. After a thorough warm-up, a few pulls were made. This simple project picked up a solid 20 rear wheel horsepower and 13 ft.-lbs. of torque. Both were up across the board, from 3500 rpm all the way to the rev-limiter.

Roller rockers such as these from T&D Machine add a rollerized tip for less friction, which creates less heat and frees up some additional power.

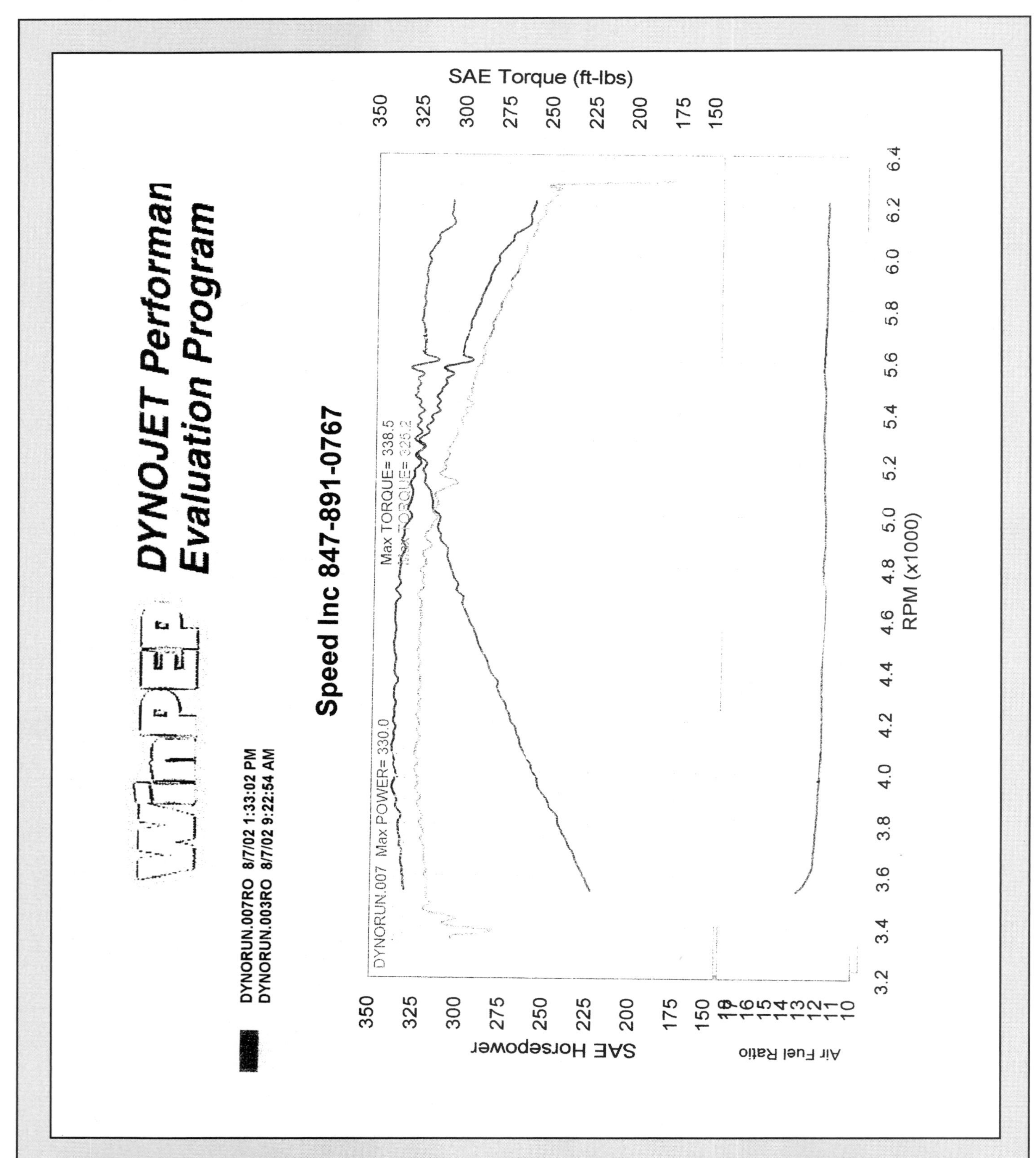

Dynamometer testing shows increased power and torque across the entire rpm range after the install of the SLP 1.85 rocker arm package.

Jesel Roller Rocker Modification and Installation

The ultimate adjustable roller rocker system for the LS1 has to be the Jesel J2K system. These shaft rocker arms systems feature tool steel lash adjusters in an ultra-light .750-inch aluminum body. The rocker body is shot-peened, equipped with a .250-inch needle-nose roller and available in a variety of ratios. The innovative .375-inch shaft eliminates the need for spacers or snap rings. Jesel includes a full complement of ARP fasteners with their J2K systems. They are available in a variety of ratios, Agostino Racing Engines elected to stay with the standard 1.7 ratio for this engine. They walked us through their installation the 418 super stroker.

They hit a snag when they began to bolt on the rocker stands. The stands were not designed with the huge 1.550-inch springs in mind, and interfered to the point that the stands could not be bolted down. As you can see, ARE milled a "pocket" for each valve spring. This is not an issue in applications with stock size springs.

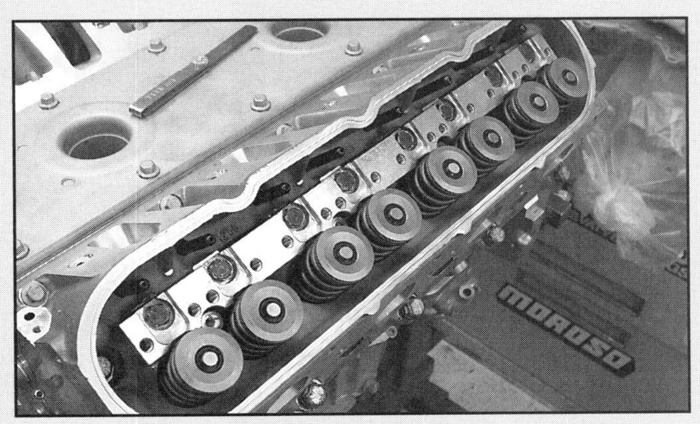

The rocker stands could now be torqued to the specified 24 lb.-ft. With that minor problem passed, the rest of the conversion should have been smooth sailing. Wrong. The bottom of the Jesel rocker arms interfered with the valve spring retainers. Again, this was a result of the huge 1.55-inch valve springs, and is not an issue with stock sized coils.

Here is a shot of the underside of the unmodified rockers.

Here is how the rockers looked after a visit to the ARE porting room for a little clearance work. While a substantial amount of material had to be removed, Agostino asserts that this poses no durability issues over the life of the rockers.

Back on track again, it was time to bolt down the rockers. The threads were coated with thread locker before being installed and torqued to 38 lb.-ft. Loctite® is always a good idea on parts subjected to extreme vibration. Valve lash was set at .020-inches on the intake and .024-inches on the exhaust.

Before the heads were torqued down on this 418 stroker, the piston-to-valve clearance was checked. To do this, a head gasket was put in place and the valves were assembled with very light checking springs. Piston-to-valve clearance was a generous .140 on the intake and .165-inches on the exhaust, thanks to the valve reliefs on the Ross pistons.

The LS1 uses light-duty beehive-style valve springs.

Piston-to-Valve Clearance

Because aggressive camshafts open the valves sooner and higher, plus keep them open longer, there is a greater risk of piston-to-valve interference. Piston-to-valve interference is a very serious problem that can result in bent pushrods, bent valves, broken rockers and even broken pistons. Always verify piston-to-valve clearance when making valvetrain alterations. Minimum piston-to-valve clearance should be no less than .080-inches.

Valve Springs

LS1 valve springs are a beehive-type constructed from 4.6mm round super clean Cr-Si wire. The reduced-diameter end coils allow a smaller, lightweight steel retainer to be used with a single bead lock. The reduced overall mass allowed the valve spring seat pressure to be decreased, thereby reducing friction and valvetrain noise.

Valve Float

Valve float is a condition where the valve spring can no longer control the action of the valve. There are a number of factors that can cause valve float such as weak valve springs, a heavy valvetrain, or excessive engine speed.

A spring that is sufficient at 6200 rpm may not be sufficient at 6300 rpm and might leave the valve hanging open when the spring should be pushing it closed. Damage resulting from valve float can range from mild to extensive. The immediate result of valve float is a radical decrease in power. At a minimum, the springs are usually permanently damaged. During valve float, part of the spring is attempting to compress while another portion is trying to expand. This superheats the spring causing it to lose its temper and decreasing spring tension.

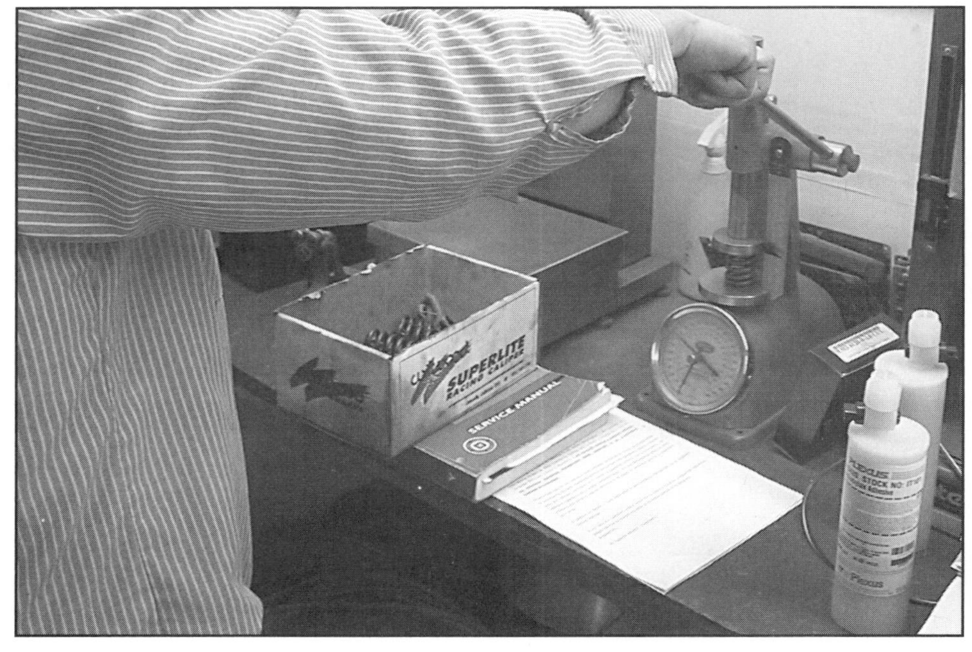

It's a good idea to verify valve spring pressure before assembly.

Dual valve springs with dampers are ideal for aggressive solid-roller combinations.

In order to accept those big springs, the cylinder head spring seat must be milled.

Hardened spring seats are designed to protect the soft aluminum head from high spring pressure.

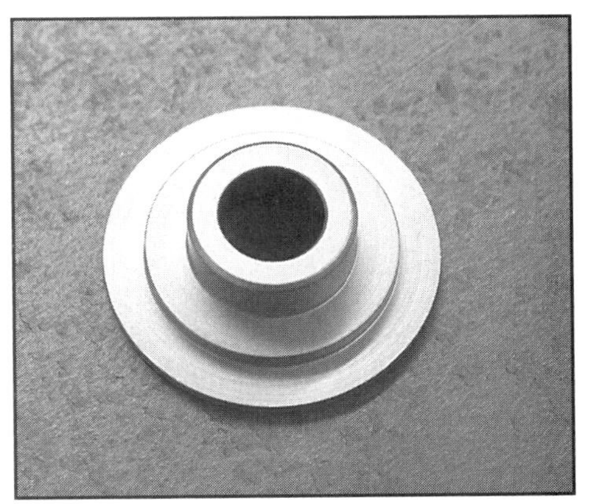

Valvetrain weight can be reduced by the use of titanium spring retainers.

Once this occurs, the valve springs are usually damaged so badly that an engine that once was easily capable of 6000 rpm can no longer reach that speed. There are other factors that can contribute to an engine that won't rpm, but if the engine has been over-revved recently and does not respond to tuning changes, chances are good that the springs are the culprit.

Choosing the Right Springs

It is imperative that the valve springs keep the lifters in contact with the camshaft lobes at all times, or valve float will occur. High valve spring pressure helps combat the violent profile changes of the lobes on very aggressive cams.

Aftermarket, beehive performance springs are available from such companies as Crane, Comp and Lunati, and will typically work well with up to about .625-inches of lift. When replacing valve springs, always consult the cam manufacturer for their recommendations.

For very aggressive cams with more than .600 inches of lift, a good option is to use 1.550-inch diameter valve springs. Springs of this size are normally utilized in big-block applications, making them readily available in countless heights and seat pressures. They are not, however, a drop-in replacement for the LS1, as they require the spring seats on the heads to be milled. The most beefy springs of this type are best suited for solid rollers, because with over 200 lbs. of seat pressure and 600 lbs. of open pressure being the norm, they could easily collapse a hydraulic lifter.

Retainers

The spring retainer works in concert with the locks to secure the spring on the valve stem. The retainer's job is simple: prevent the valve spring from shooting off the valve. Be sure the retainers are the proper diameter to keep the valve springs from walking, and are designed to accept standard 10° locks.

Titanium—Weight plays a key role in valvetrain dynamics. Any mass added to the valve, spring, or retainer becomes a force that resists changing direction, especially at high rpm. The lighter the valvetrain components, the quicker the engine will rev. Titanium retainers are approximately 40% lighter than steel, but do not sacrifice any reliability, making them a great

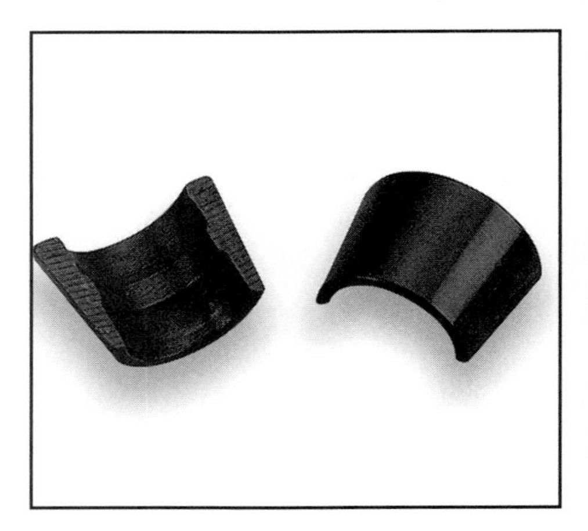

Valve locks are not a glamorous engine component, but a failure will ruin your day, and perhaps your engine.

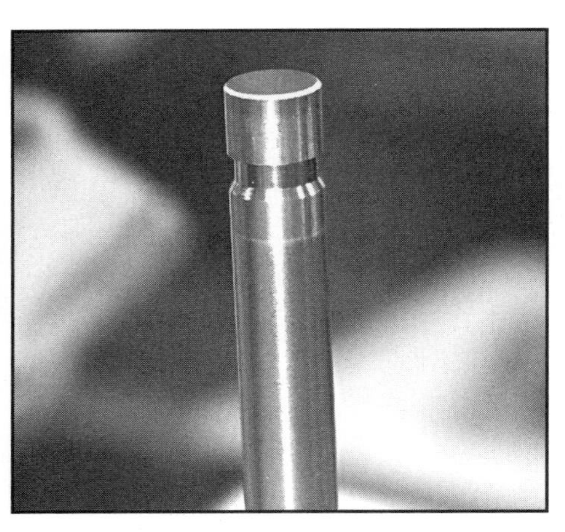

The locks seat in the machined groove of the valve stem tip.

Lash caps are installed over the tips of the valve stems, allowing the use of an otherwise too-short valve.

A rev-kit can extend the rpm range of a hydraulic valvetrain to 7000 or more. This is installed in the lifter gallery in place of the plastic lifter retainers.

upgrade for higher-revving engines. Shaving a few grams off the end of the valve is enough to add a few hundred rpm before the engine reaches valve float. `

Locks—Valve locks fit between the valve stem and spring retainer, wedging itself in place to keep the assembly together. The small ridge on the inside of the lock positively locates it on the valve stem, while the spring retainer's tapered inner diameter is positioned over the lock's tapered outer diameter to keep everything in one place. Due to their superior strength, machined locks are vastly superior to stamped versions for not much more money. Be sure the locks are 10° to match the retainers.

Rev Kits

Because a lifter is heavy, it has a great deal of inertia, and therefore can be difficult to keep it in contact with the camshaft lobe at high rpm. The traditional means of addressing this problem is through increased valve spring pressure. While this is fine for solid cam setups, too much spring pressure can collapse hydraulic lifters causing bent or broken pushrods, and that can be catastrophic. While there are other, more exotic methods in existence to address this problem, most of them (e.g., titanium lifters) are too costly for a street engine. Minimizing or eliminating valve float also extends the engine's power band. Most engines show a dramatic power loss as the rpm continues to climb past the power peak.

A rev kit allows for higher engine operating speeds by applying spring pressure directly to the lifter to keep it in contact with the cam lobe, thereby reducing the likelihood of valve float. This becomes most critical for the LS1 operating above 6500 rpm for extended periods. Agostino Racing Engines produces a rev kit that is fitted beneath the cylinder head in the block's lifter gallery, eliminating the lifter retainers. The kit consists of twelve CNC-machined billet aluminum guide plates and a set of

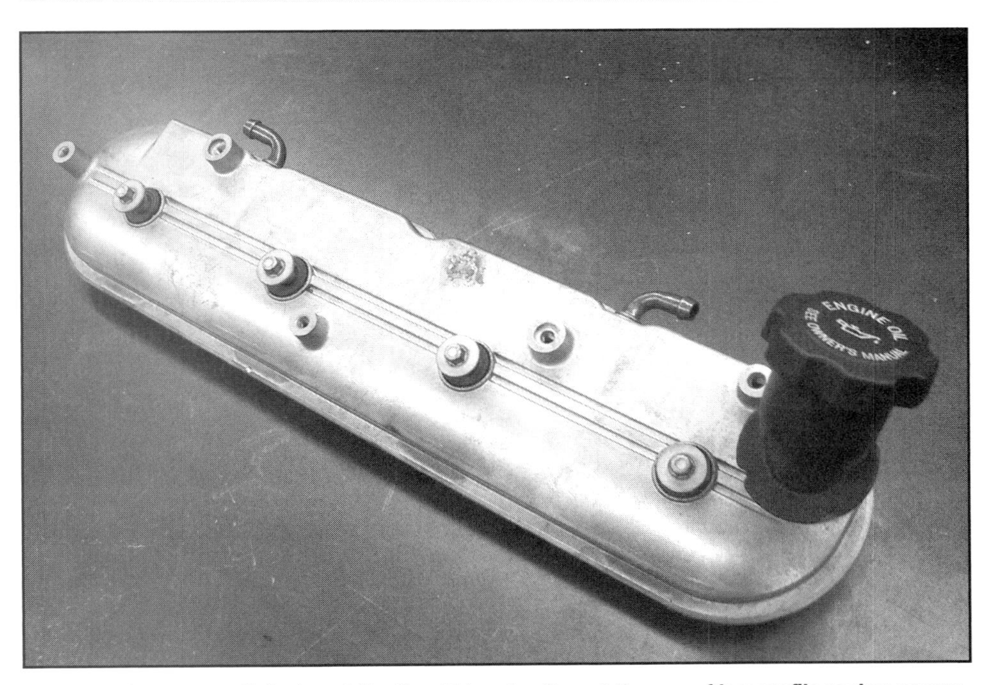

The tall valve cover rail design of the Gen-III heads allowed the use of low profile rocker covers.

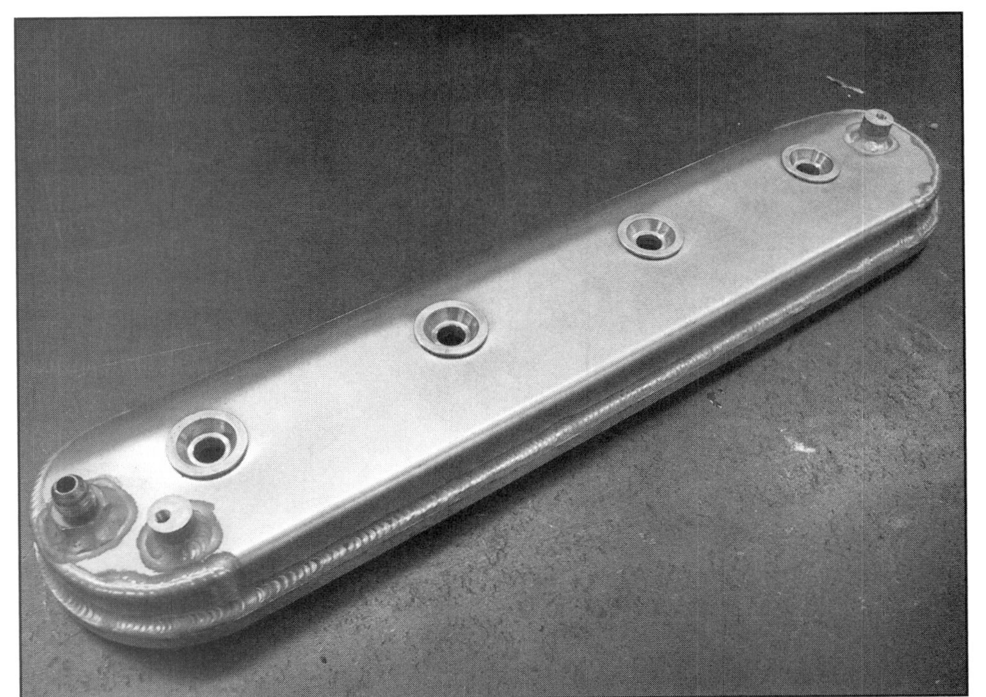

Fabricated aluminum valve covers are an elegant solution to rocker arm clearance issues.

sixteen small-diameter valve springs. While the kit does create a small amount of valvetrain noise, it is much less than the clatter of solid lifters. It is a good life insurance policy for your engine that will allow it to rev reliably and valve-float-free to 7000+ rpm. Rev-kits are best suited to road race applications that see extended high-rpm usage. This extension of the engine's rpm range allows the use of a more aggressive camshaft that shifts the power band higher. However, the elimination of the lifter retainers means the cylinder heads will have to be removed to make a camshaft change.

Valve Covers

The Gen-III valve covers are cast aluminum and use a pre-molded silicone gasket for sealing. Incorporated into the covers are mounting provisions for ignition coil-pack mounting brackets, the oil fill tube, the Positive Crankcase Ventilation (PCV) system, and engine fresh air passages. The cylinder head's tall rails allow the valve covers to be low profile. This creates a problem when moving to adjustable roller rockers, as there is just not enough space to fit them under the short covers.

There are two potential solutions to this problem. The first is a valve cover spacer, which will allow you to retain the stock covers. Spacers are inexpensive, but will require longer plug wires and are more prone to leakage. The other option is an aftermarket valve cover, such as GM Performance Parts' C5-R covers. These are designed for use with Jesel or T&D roller rockers, and are significantly taller than stock. They will bolt right on to '99+ cylinder heads with center bolts, but do present a couple of other small obstacles. First is the lack of any gasket provisions. Having an O-ring groove machined into the top of the cylinder head rail can solve this. While it looks rather delicate, it seals perfectly.

The C5-R valve covers do not have coil-pack mounting provisions as the OEM valve covers do. This is not necessarily a bad thing, as relocating

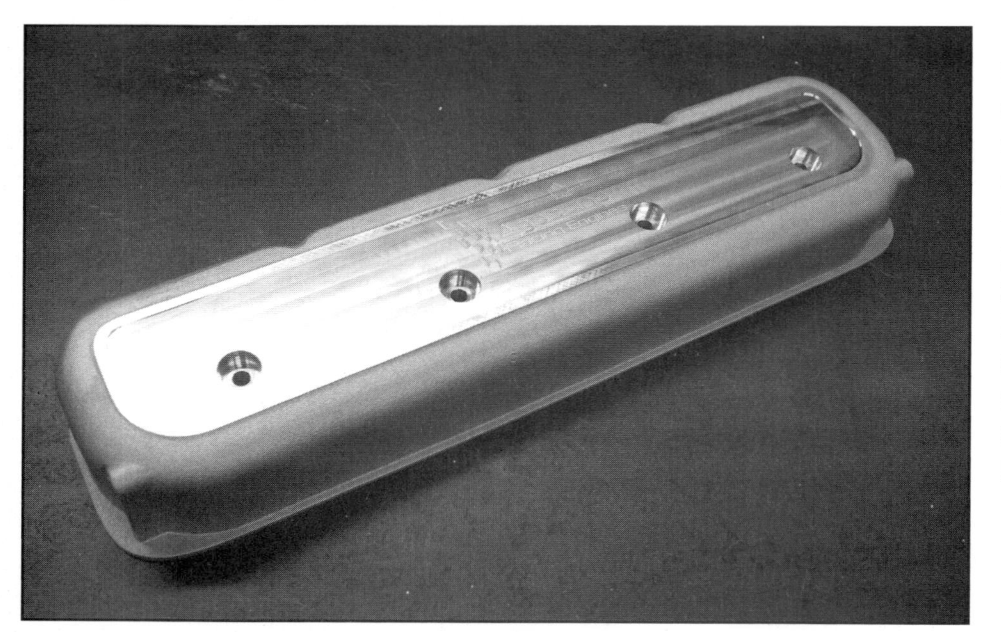

GM Performance Parts C5-R valve covers bolt right on '99 and later heads and provide plenty of clearance for aftermarket roller rockers.

them makes it easier to remove the covers for lash adjustments if you are running a solid cam. One option is to fabricate a custom bracket to mount the coils closer to the fuel rails. Or, if you are building a racecar, you can mount the coils in the former locations of the air-conditioning accumulator and ABS module. With some of the clutter removed, the LS1 can actually be a pretty darned good-looking engine!

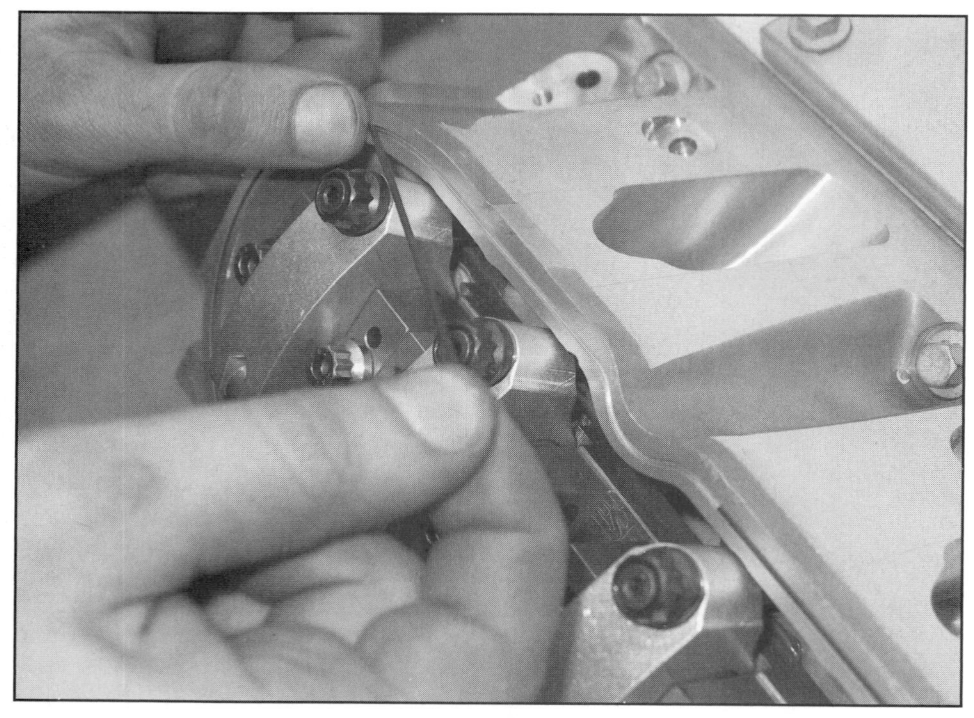

Because C5-R valve covers have no gasket provisions, other arrangements were necessary. An O-ring groove can be machined into the cylinder head rail for valve cover sealing.

8 Exhaust Systems

Scavenging ability is the key factor by which an exhaust system should be evaluated. The more efficient the system, the more air/fuel mixture can be pulled into the combustion chamber. Automakers often have to compromise performance for other manufacturing considerations, such as ease of production and reduced cost. A high-flow exhaust system, properly sized for your engine should be a top priority. The right exhaust system will give your car improved power, better fuel economy and a great sound.

Technical Data

LS1 exhaust manifolds are constructed with two walls of stainless steel. The 309-series inner wall is .8mm thick and 409-series outer wall is 1.8mm thick. There is a 3mm insulating air gap between the inner and outer walls that preserves heat in the exhaust stream, ensuring rapid converter light-off and reducing cold start emissions. Cast iron exhaust manifolds replaced the fabricated manifolds in the 2000 model year.

Oxygen sensor provisions are located in the takedown pipes of each manifold. Each takedown pipe feeds an under-floor catalytic converter that is located 600mm downstream of the takedown flange. The monolith converters have an oval cross-section and displace 1.39 liters each. OBD-II monitors catalyst function by comparing data from a second pair of oxygen sensors downstream of the converters. When the downstream sensor data mimics the upstream readings, the PCM interprets a degradation of converter performance and lights the Service Engine Soon lamp.

The electric Air Injection Reactor (AIR) pump runs after start-up for a short period of time dictated by the PCM. An integral runner injects air from the AIR system into each manifold at the head flange lessening start-up emissions. The Corvette does not require the use of Exhaust Gas Recirculation (EGR), though F-bodies utilized an EGR system through model year 2000. The introduction of the LS6 manifold, plus improved catalytic converters, allowed elimination of the EGR system in 2001.

LS1 exhaust systems are equipped with four oxygen sensors, two upstream and two downstream of the catalytic converters.

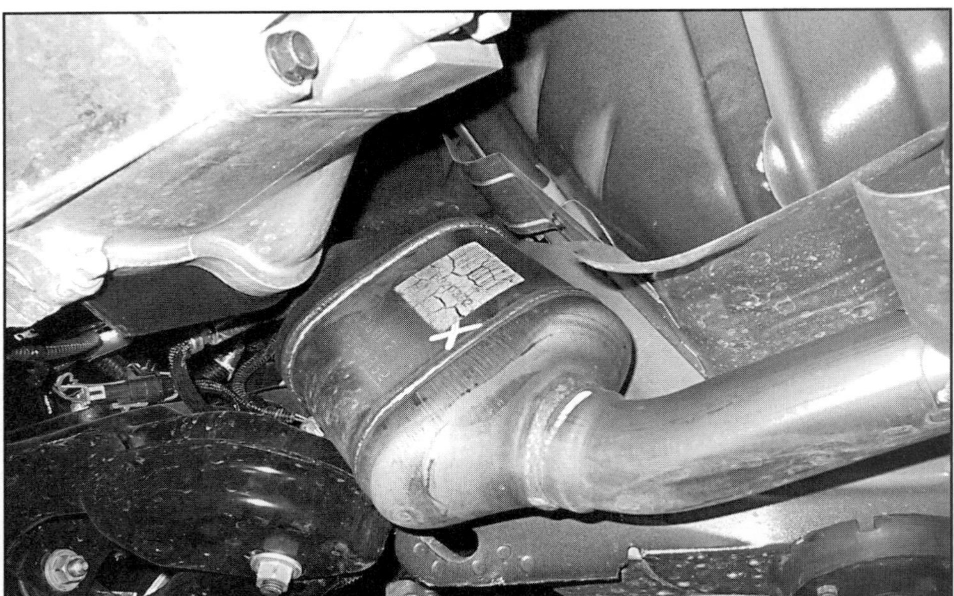

The stock catalytic converters performance is monitored by the car's PCM through the oxygen sensors.

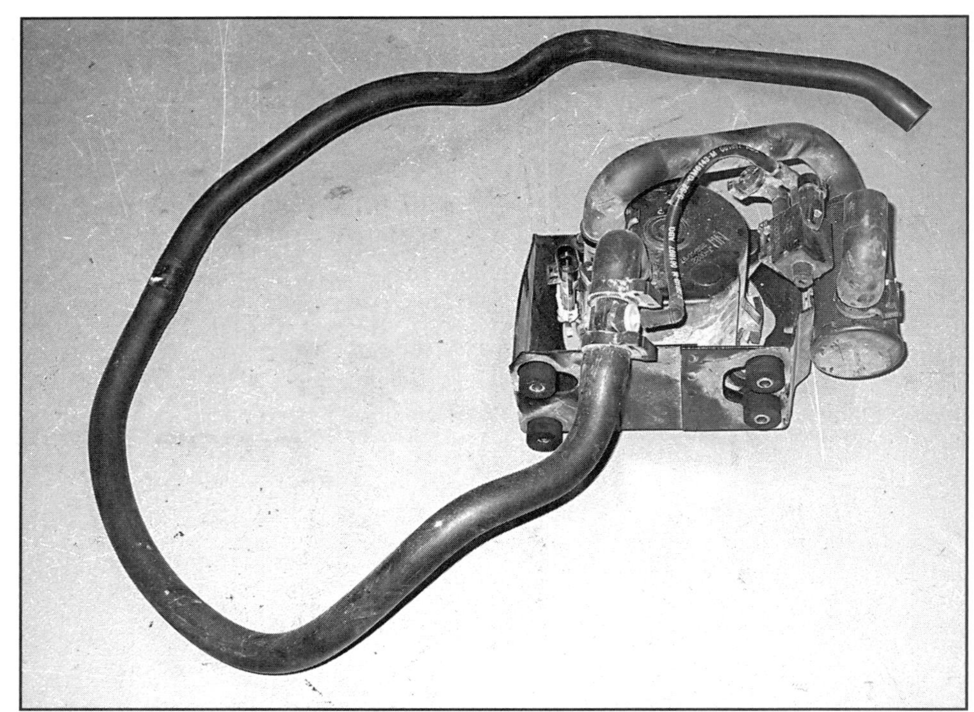

The LS1's AIR pump system lessens engine emissions during start-up.

Exhaust manifolds for all years are sealed to the cylinder heads with a three-piece gasket. The manifolds are fastened to the cylinder head with six nonreusable bolts. The bolts are coated with a pre-applied adhesive thread patch to permit thermal expansion and contraction of the manifold flange while maintaining proper fastener tension.

The balance of the F-body exhaust consists of a Y-pipe that merges exhaust streams from both banks and moves it into 2.75-inch intermediate pipe and onward to the muffler. The transverse-mounted muffler is located near the aft end of the car, beneath the fuel tank. Depending on the year and model of the car, either dual 2.25 or single 3-inch tailpipes were installed.

The Corvette benefits from true dual exhaust with 2.25-inch pipes and an H-pipe aft of the catalytic converters. Dual mufflers are mounted transversely at the rear of the car and each is fitted with two tailpipes. The Z06 uses a lightweight titanium version of this system with the same layout.

Performance Exhaust Systems

There are a myriad of headers, X-pipes, Y-pipes, and cat-back exhaust systems on the market for both the F-body and the Corvette. One of the keys to optimizing performance is matching the right combination of these components to your engine.

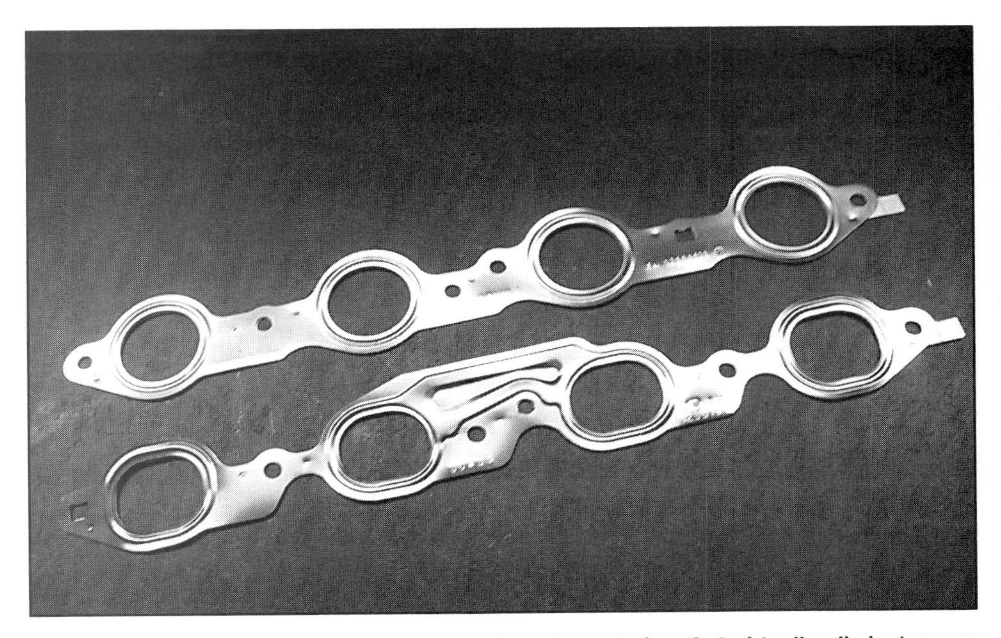

The stock exhaust manifold gaskets are a three-piece design that virtually eliminates any leakage problems–even with headers.

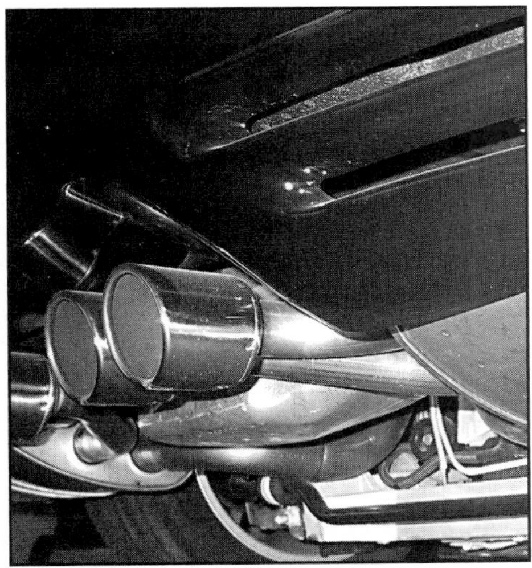

Corvette Z06 exhaust sports lightweight titanium mufflers and tailpipes.

F-body exhaust systems utilize a cross-flow muffler with dual tailpipes.

Tubing Diameter

A critical decision that needs to be made when selecting an exhaust system is tubing diameter. It is important not only that the pipe diameter be properly sized for the application, but that it remains constant throughout the length of the system. Changing the tubing diameter downstream will restrict smooth exhaust flow by creating turbulence and excessive backpressure. Anything that causes turbulence or backpressure in your exhaust system reduces the power that engine modifications can create.

Headers

It's no secret that a good pair of headers are an excellent performance upgrade. While the stock exhaust manifolds perform admirably, they do create substantial exhaust gas turbulence because the primaries are very short and dump directly into a single tube. Headers are more efficient because they reduce back-pressure and keep each cylinder's exhaust pulse separate for a longer time. This reduces the possibility of spent gases flowing back into an adjacent cylinder and contaminating the incoming charge mixture, particularly when using a cam with increased overlap. Headers will usually give a demonstrable, but not huge performance gain even when connected to an otherwise stock exhaust system.

Header Size—It's natural to assume the bigger the header tube the better, but that's not the case. The truth is that overly large primaries actually cost torque and horsepower because they cause a decrease in exhaust gas velocity. Smaller diameter pipes will ultimately flow less volume than large

Headers are a great way to free up some horsepower on any engine.

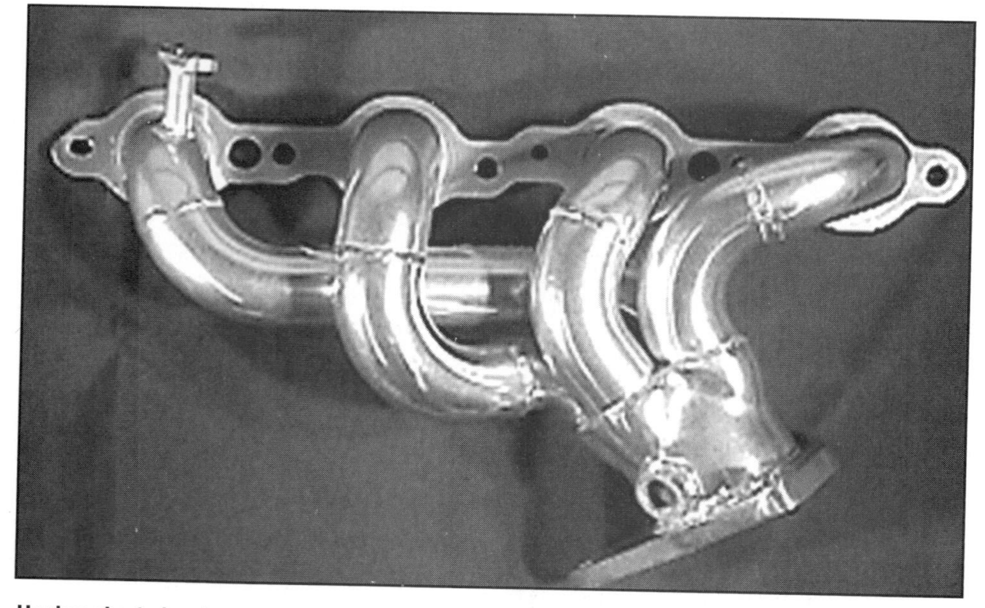

Hooker shorty headers are a great addition for mild performance build-ups.

ones, but until you reach the RPM where the sheer volume of exhaust gases require bigger primaries, smaller tubes scavenge more efficiently.

Unless you do most of your driving at full throttle, you'll be better off sizing your headers conservatively, even if you're running a radical cam or a blower. If you're using the engine mostly in the idle to 6000-rpm range, which is typical for a street-driven car, 1.625-inch primary tubes are a good choice. Larger tubes will cause a small decrease in low-end torque. Small tubes usually don't begin to lose their edge in horsepower or torque until about 5000 rpm.

Shorty Headers—For a car that rarely sees the racetrack, 1.625-inch shorty headers may be worth considering. Shorty headers utilize short primary tubes and locate the collector in the place of the stock manifold's downstream flange. This means they are able to replace the factory manifolds with slightly better flowing tubular weldments that allow the retention of the stock Y-pipe and emissions equipment. Hooker and JBA "Cat 4-ward" shorty headers allow the reuse of the stock Y-pipe on F-bodies.

Mid-Length Headers—MAC produces what they term a "mid-length" header that is a very good performance option. They have 1.75-inch primary tubes and a 2.5-inch collector. These are available with or without a Y-pipe equipped with catalytic converters. The downside of this system is that it requires use of MAC's Y-pipe. Gains with this system can be as much as 15 rear wheel horsepower.

Long-Tube Headers—Full length, or long-tube headers use very long primary tubes that position the collector under the vehicle's floorboards. Long tubes are commonly available in two primary tubing diameters: 1.75- or 1.875-inch. Most are available with 3-inch diameter collectors, though 3.5-inch collectors are not unheard of. SLP, Hooker, Jim Pace Performance Parts and Finish Line Performance are but a few of the companies offering top-quality full-length headers.

Equal-Length Headers—Headers with equal-length primary tubes have been shown to develop slightly more power in an open, unmuffled system, that advantage is quickly negated when catalytic converters, mufflers and tail pipes are bolted up behind them. In order to take maximum

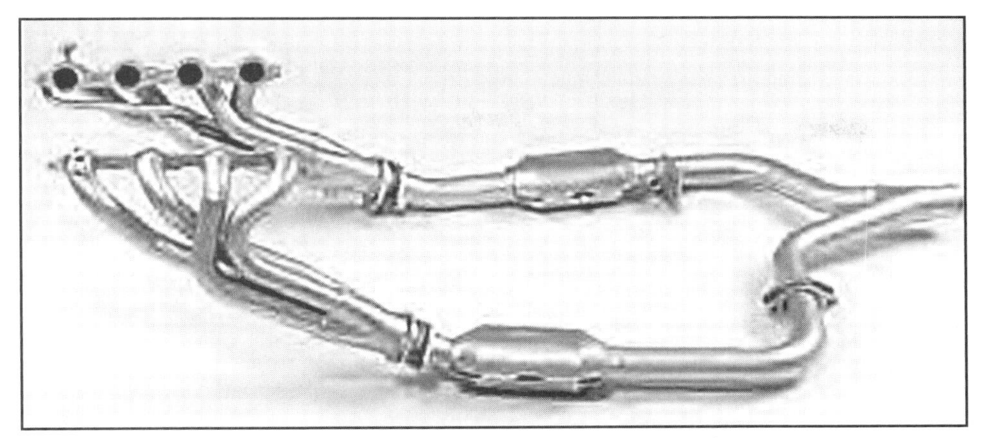

MAC mid-length headers are a good performance option for street/strip cars.

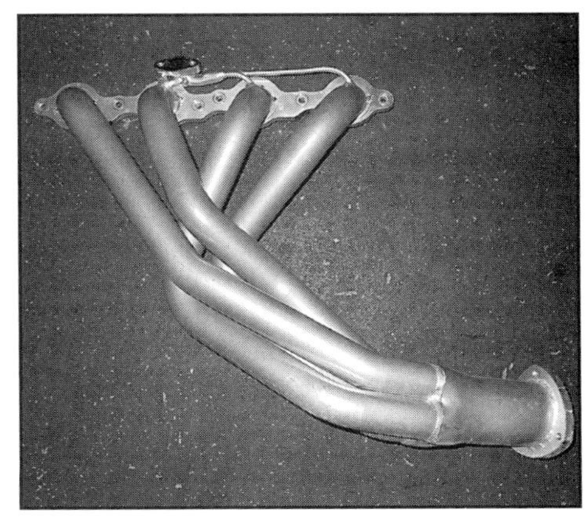

Long tube headers are the only way to go when maximum power is required.

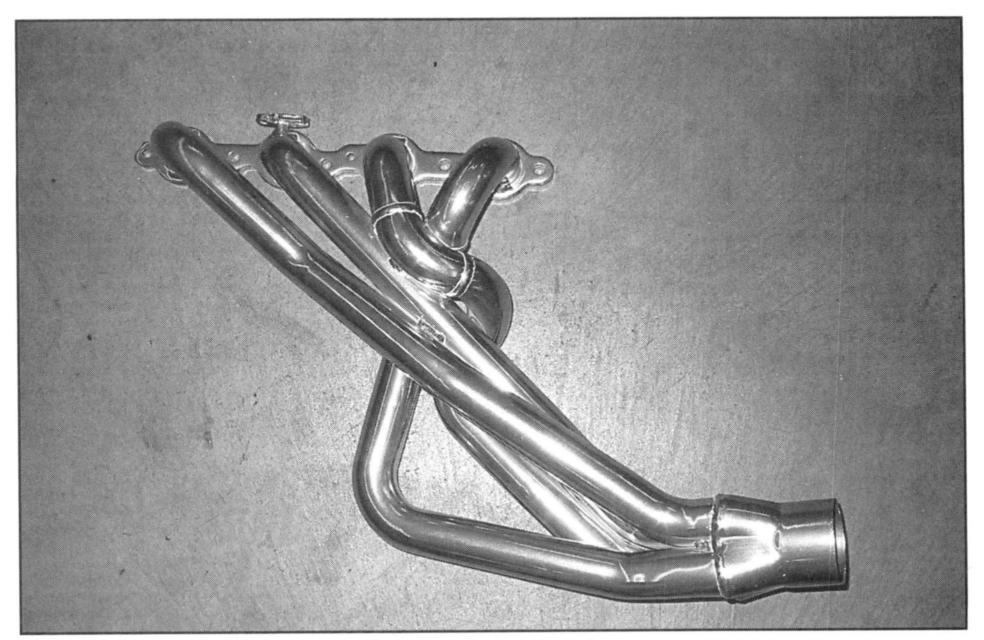

SLP's are the first full-length headers made with tuned tube lengths for LS1-powered F-bodies. The length of each header tube was calculated to maximize scavenging, which translates to faster revs and increased torque and horsepower. SLP's headers utilize stainless steel 1-3/4" primary tubes with a 3" stainless steel collector and a high-luster thermal coating.

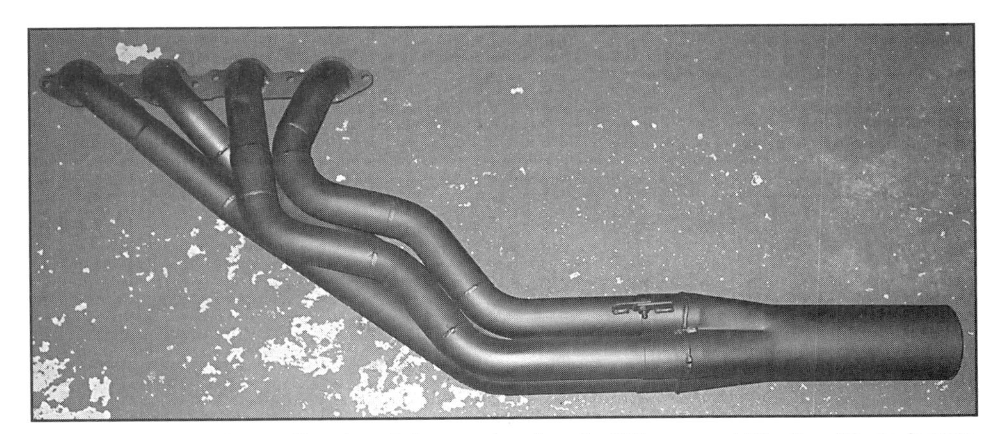

Pace Performance Parts offers long-tube race headers for LS1-powered F-bodies. These feature 1.75 to 1.875 -inch stepped primaries with adjustable 3.5-inch collectors.

advantage of equal length tubes ability to produce more power, their length needs to be perfectly matched for the engine at the rpm range where maximum power is desired. What that means is a lot of custom fabrication and dyno testing, which translates directly into a lot of time and dollars spent. For an all-out racecar, the gains may justify the expense, but street cars are much better off with a production header. While they may not perform optimally with an exhaust system installed, equal length headers won't hurt performance.

Stepped Headers—Stepped headers are long-tube headers with one important difference: they have primary tubes made of two or three different diameters of tubing. The smaller diameter tubing emerges from the header flange and larger tubing is slipped over and welded at a calculated distance downstream. The change in diameter is called a step. Most stepped headers consist of either two (single step) or three-tube diameters (double step). The idea behind this design is to create a venturi effect that increases exhaust gas velocity. As with equal length headers, much fabrication and dyno testing is required to maximize the effectiveness of step headers, making

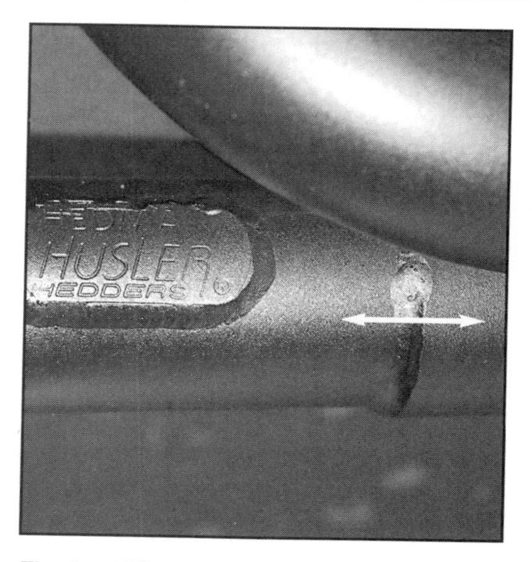

The transition between tubing diameters is known as a "step."

these practical only for maximum-effort racecars.

Merge Collectors—Merge collectors replace the standard collectors on a pair of headers and can significantly improve the efficiency of the exhaust system by minimizing the change in cross-sectional area where the primary tube transitions to the collector. This is done in an effort to maintain as much exhaust gas velocity as possible. The ability of a header/exhaust system to maintain exhaust gas velocity has a tremendous impact on torque and horsepower. Any significant changes in the system, such as collectors or mufflers will influence exhaust gas velocity. The substantial change in tubing volume in the transition primary tube to the collector causes a big reduction in exhaust gas velocity, which decreases the efficiency of the system and its ability to scavenge the combustion chamber.

Simply replacing standard header collectors with merge collectors seldom produces a power gain. Dyno testing is the best way to arrive at the final camshaft combination when using a merge collector. If you need

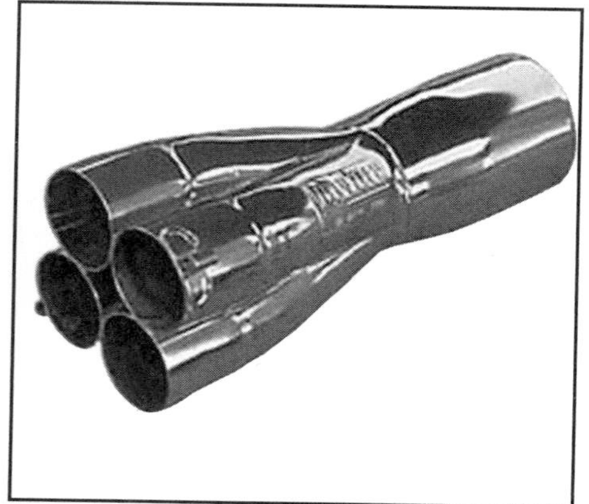

Merge collectors, when configured properly, can significantly increase engine power. Unfortunately, much testing is required to maximize the gains.

that last little edge and are willing to do the testing required to realize the potential gains, merge collectors are a good way to gain significant torque and horsepower.

X-Pipes and Y-Pipes

After the exhaust gasses pass through the primaries and collectors of the headers, they enter the car's X- or Y-pipe. It is these components that hold the catalytic converters (if so equipped) and transfer the gases on to the mufflers and tailpipes. Most header manufacturers offer Y-pipes (F-body) or X-pipes (Corvette) to match their headers. These are usually offered in emissions-legal versions equipped with high-flow catalytic converters or off-road versions where the cats are eliminated. Because header designs vary significantly, it is a good idea to buy the headers and Y- or X-pipe as a set, or invest in a custom pipe tailored to your specific needs.

Catalytic Converters

When catalytic converters started appearing on cars in the mid-1970's, they were immediately scorned by performance enthusiasts for the severe chokehold they placed on the exhaust systems. It was standard practice to remove the converters and replace

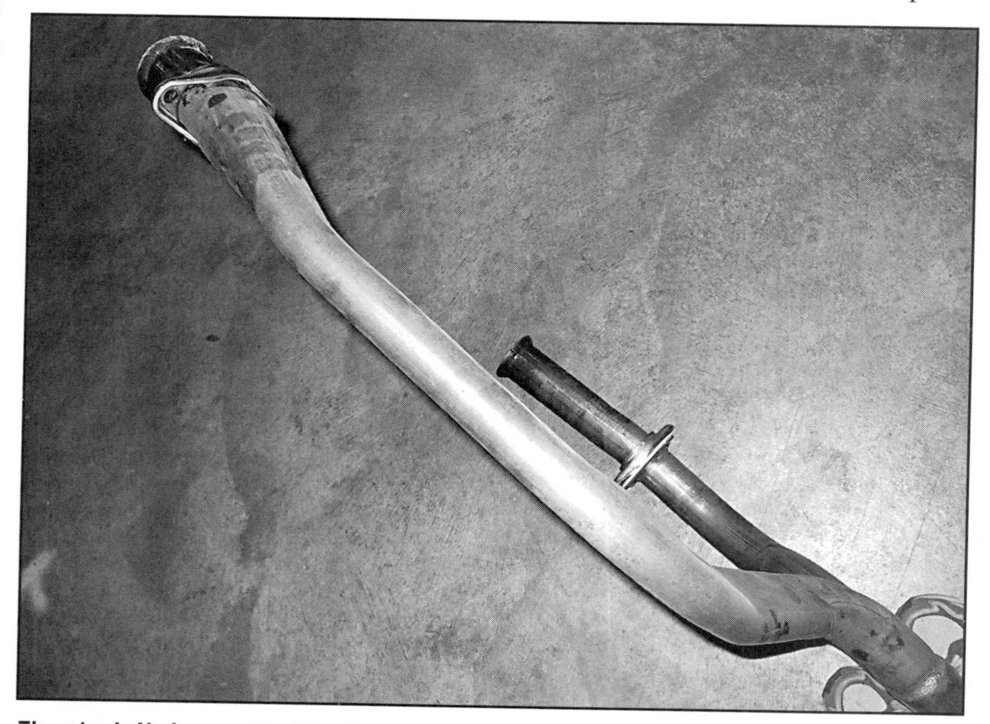

The stock Y-pipe on the F-bodies exhibits this flattened section on the driver's side. Theoretically, it should improve ground clearance. Realistically, it is a flow restriction.

SLP's stainless steel Y-pipe for their full-length headers accommodate two Random Technologies high-flow catalytic converters.

O2 sims are plugged in to the harness in place of the oxygen sensors. Their purpose is to generate a signal the PCM will see as normal.

them with a standard length of exhaust tubing, commonly known as a "test pipe." While this was (and still is) illegal, enthusiasts were willing to risk the possible legal consequences in the name of performance.

As with all other aspects of automobile manufacturing, catalytic converter technology has improved by leaps and bounds since the dark performance days of the '70s. Nowadays, stock catalytic converters reduce exhaust flow only marginally. They are best left alone until long-tube headers are installed. Even then, a pair of high-flow aftermarket catalytic converters such as those from Random Technologies will lessen engine power by no more than 2 or 3 horsepower compared to an exhaust system without converters. The small power increase is generally not worth the hassles involved with making the car "street legal" and capable of passing an emissions inspection. The only real advantage to removing the cats is that you can then use leaded race fuel, though leaded gas will wreak havoc with your oxygen sensors whether the cats are in place or not.

Oxygen Sensor Simulators

If you elect to eliminate the catalytic converters from your exhaust system (which, by the way is against the law, so do so only for off-highway use or at your own risk!), you will be faced with a bright yellow "Service Engine Soon" light. This can be addressed by installing oxygen sensor simulators. O2 sims are simple electronic devices that plug in the harness in place of the rear O2 sensors and generate a random voltage to keep the PCM happy. The other option is to have a PCM programmer disable rear O2 sensor functionality. Both methods work equally well. O2 sims are available from suppliers such as LS1 Motorsports and Thunder Racing.

Exhaust Cut-Outs

Cut-outs are an economical way to run open exhaust without the hassle of exhaust system removal. These are typically spliced into the exhaust either in place of the catalytic converter, or between the cat and the muffler. An open cut-out reduces the restriction of the cat-back system, but makes for an extremely loud ride. If you wish to maximize your car's performance at the track by running open exhaust, you are better off removing the entire exhaust system aft of the headers. Not only will this remove all airflow restriction, but you will also gain the additional benefit of a 60-pound (approximate) weight savings.

Cat-Back Exhaust Systems

Cat-back exhaust systems, so named because the generally replace all components aft of the catalytic converters, are one of the most popular modifications for all LS1-powered cars. Though they have a rather sedate exhaust note that leaves most enthusiasts wanting more, the factory exhaust systems flow very well. This is not an area to go in search of big power, as typical gains are no more than 10 rwhp. If you decide to install a cat-back, do it for the looks and the sound. Some manufacturers even have sound bites of their systems available for download on their websites. The majority of cat-back exhaust systems

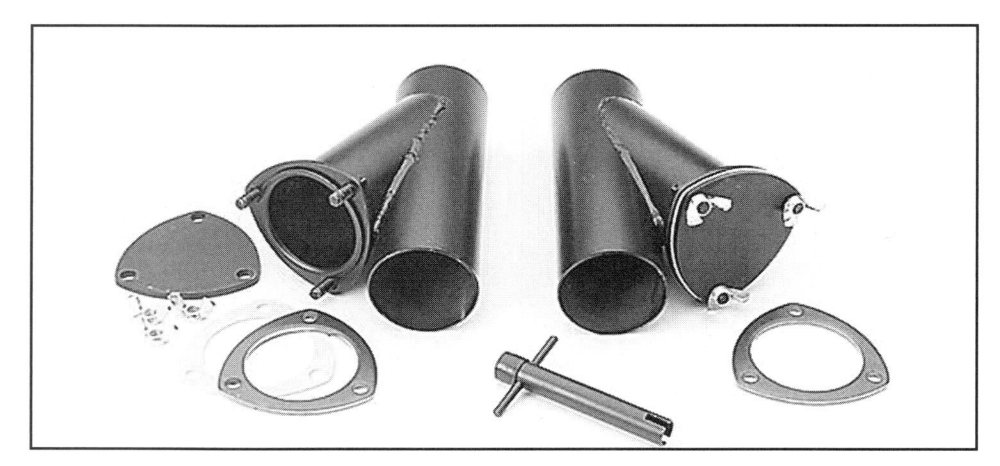

The sound of an open cutout will definitely turn heads.

Performance cat-back exhaust systems for the F-body, like this one from SLP, usually add about 10 rear wheel horsepower to an otherwise stock car.

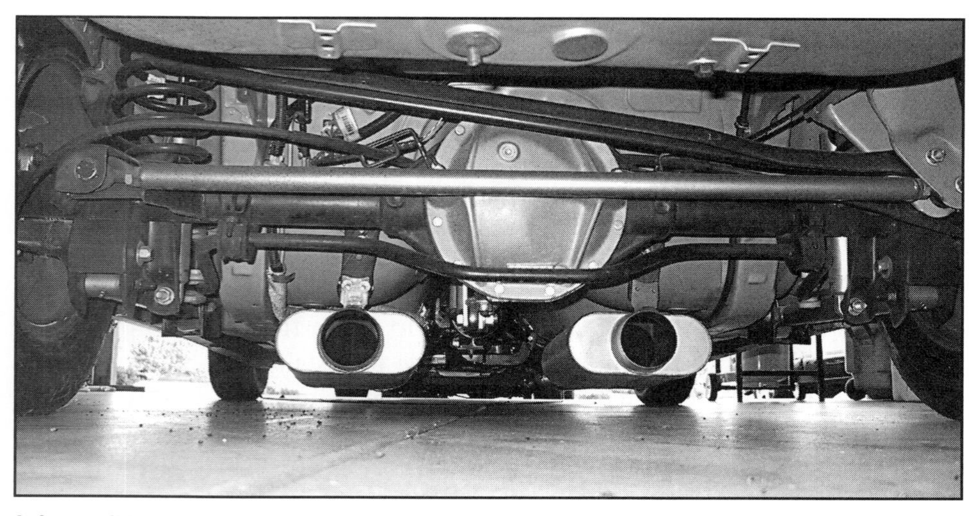

It is possible to build a dual exhaust system for an F-body, but it is largely unnecessary given the high-quality cat-backs on the market.

are emissions legal.

F-body Options—Most cat-backs for the F-body utilize a 3-inch intermediate pipe, a stock location, low restriction muffler and large single or dual tailpipes. Single tailpipe systems are slightly more efficient because they create less turbulence inside the muffler. Make your choice based on whether you prefer the dual exhaust look. Most cat-backs are constructed from stainless steel, though some budget systems are aluminized mild steel. The best cat-backs for the buck are those from Flowmaster and Hooker. Other companies, such as Borla and B&B Fabrications, offer meticulously fabricated stainless steel systems. For the truly hardcore, Mufflex offers a 4-inch diameter single tailpipe system.

Dual Exhaust—True dual exhaust systems for the F-body are possible, but these often compromise ground clearance. There are currently no systems of this type available, so this is a custom-only proposition best suited for racecars.

Corvette Options

Aftermarket Corvette systems typically consist of a rear muffler section that bolts up behind the factory H-pipe. These place the mufflers in the stock position, mounted transversely behind the rear tires. Consider the installation of an X-pipe to replace the factory H-pipe for a small power increase and noticeable decrease in interior resonance.

As with F-body systems, the vast majority of Corvette cat-back systems are offered in stainless steel, although a few companies like Borla and Corsa are now building titanium systems. Top-quality Corvette systems are

B&B Fabrications offers several cat-back systems for the Corvette, all constructed of polished stainless steel.

available from Borla, B&B and Corsa. Though budget and Corvette are not often used in the same sentence, check out the offerings from Dynomax or Flowmaster if you wish to save a buck or two.

Central Muffler Systems

Central muffler systems are gaining popularity because of decreased weight and the reduced tendency for heat to warp the rear bumper cover. The disadvantage of a central muffler system is a very loud exhaust note and quite a lot of interior resonance. Borla and Corsa are two popular choices for central muffler systems.

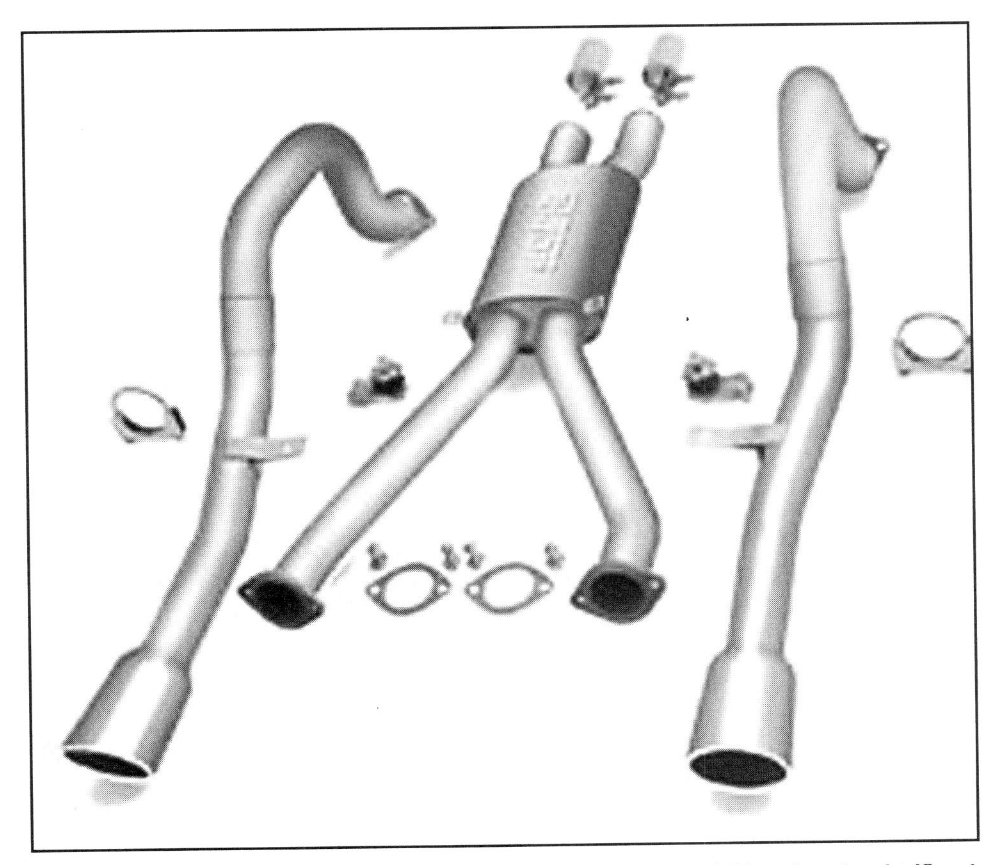

Central muffler systems like this one from Borla trade significant weight savings for significant muffling of the exhaust note. These are LOUD!

9

Fuel System

The fuel system is an area often overlooked by many in pursuit of their performance goals. The importance of the fuel system cannot be overstated: Ensuring sufficient fuel supply is critical for maximum performance and good driveability. The fuel flow capacity must be matched to the airflow requirements of the engine, as one without the other will not make power and may negatively impact the life of your engine. Your fuel system should meet the demands of your engine and its attendant modifications, both in volume and pressure.

The key components of the LS1 fuel system are eight injectors, a fuel rail, a .375-inch supply line and an in-tank electric fuel pump. The LS1 utilizes a nontraditional, semi-returnless fuel system architecture to meet evaporative emission requirements. Fuel is regulated to 58.02 psi (4-bar) by the in-tank regulator, which ensures that all fuel passing into the rail is consumed by the engine. The fuel rail is an H-style fabricated stainless steel piece incorporating a quick connect fitting for fuel line attachment. The fuel system on the 1997 Corvette differed slightly; fuel pressure was regulated at the fuel rail entrance and manifold vacuum referenced.

Fuel Injector Overview

GM has installed various injectors with differing flow rates in the LS1 over the years, ranging from 26.4 to 29.1 lbs.-hr, rated at the LS1's 58.02 psi fuel pressure. In most cases, the OEM injectors are capable of supplying enough fuel to comfortably support 400–425 rwhp. However, use of stock injectors is not recommended for service in engines subjected to extended intervals of wide-open throttle. Long-term operation at near 100% duty cycle will drastically shorten an injector's lifespan.

The amount of fuel delivered by an injector is determined by the length of time it is held open by the PCM and by the system's fuel pressure. An injector must be able to respond quickly, supply a small amount of idle fuel, and a large amount at wide-open throttle to prevent high-rpm lean-out. These wide operating boundaries demand that an injector have a broad dynamic range of operation.

Fortunately, upgrading the injectors

Stock fuel injectors can typically support up to 425 rear wheel horsepower comfortably.

is a simple procedure, but requires some forethought to ensure a good match is made for the engine modifications. Poorly matched injectors can cause driveability problems such as rough idle, surging, poor throttle response or even high-rpm lean-out, which can lead to serious engine damage. An injector change also requires a change to the PCM injector calibration tables to ensure proper fuel delivery.

Injector Types

There are 3 different types of injector nozzles suitable for the LS1:

Pintle—This is the most common type of injector, where a small tapered needle (pintle) reciprocates in and out of a matching orifice. When the injector solenoid is energized, the needle is pulled back, allowing the fuel to be discharged. Pintle injectors have a moderate spray pattern suitable for most applications and are a widely used OEM injector type.

Disc—Disc-type injectors come in two varieties from two manufacturers: Bosch and Lucas. The Bosch-type uses the same actuation mechanism as the pintle injector, but replaces the pintle with a flat disc and a plate with a pattern of holes. The Lucas-type

Fuel injector impedance can be measured easily by measuring the resistance across the harness connector on top of the injector.

moves the disc inside the body to reduce the mass of the assembly for quicker response. These typically have a very narrow spray pattern, which can affect idle and throttle response in some cases. In most instances, however, they have been found to work well in an LS1 application.

Ball—The Rochester division of GM makes ball-and-socket-type injectors for OEM applications and keep race types for MSD. These use a ball-and-socket arrangement and have excellent atomization with a wide spray pattern.

Before you can make an educated decision regarding fuel injectors for your engine, there are some terms you should understand:

Impedance—Impedance describes the electrical resistance of the solenoid windings inside the fuel injector. The simplest way to determine impedance is to measure the electrical resistance across the two electrical connections with a multimeter. This simple test will determine an impedance value, measured in Ohms. Injectors are usually grouped in to two broad categories:

Peak and Hold or Low Impedance (current mode): 2.0 to 5.0 Ω

Saturated or High Impedance (voltage mode): t 12 to 16 Ω

General Motors has used both low and high impedance injectors through the years. Currently most engines, including the LS1, use a high impedance injector. The primary advantage of high impedance injectors is that less heat is generated in the drive circuit. The largest high-impedance injector currently available is a 50 lb.-hr unit (@ 43.5 PSI) from MSD.

One of low impedance injectors' biggest advantages is a shorter triggering time, which gives better idle quality. Also, most aftermarket fuel management computers are set up for low impedance units as injectors exceeding 50 lbs-hr are commonly used.

Flow Rate

The amount of fuel an injector is capable of delivering is known as flow rate and is measured in pounds per hour (lbs.-hr.). The system pressure at which an EFI system runs directly affects the delivered fuel volume of an injector, per the pulse width (time the injector is open) as commanded by the PCM. Raising the fuel pressure increases the flow while lowering it decreases the flow, assuming the same PCM command. Most injector sets can be purchased flow tested in matched sets with a tolerance of ± 1%, or can be purchased without being tested. Factory tolerance varies among manufacturers, but variances of 5% or more

Common High Impedance Injector Flow Rates (100% Duty Cycle)

Source	Part No.	Lbs.-hr. @ 3 bar (43.51 psi)	Lbs.-hr. @4 bar (58.02psi)
1997-98 LS1	12554271	24.7	28.5
1999-2000 LS1	12555894	22.3	26.3
2001-03 LS1 / LS6	12561462	24.7	28.5
Lucas	5207011	23.9	27.6
Bosch (Buick GN)	0-280-150-218	29.8	34.4
Lucas green stripe (2-hole)	5207009	30.0	34.6
Bosch brown top (early T-bird Supercoupe)	0-280-150-756	32.0	37.0
Bosch	0-280-150-911	33.0	38.1
Bosch blue tops	0-280-150-967	36.0	41.6
MSD	2018	38.0	43.9
Bosch (4.7 ohm)	0-280-150-803	40.0	46.2
Lucas red stripe	5208008	40.1	46.3
Lucas green stripe (3-hole)	5208009	42.5	49.1
MSD	2013	50.0	57.7

Common Low Impedance Injector Flow Rates

Source	Part Number	Lbs.-hr. @ 3 bar (43.5 PSI)	Lbs.-hr @4 bar (58 PSI)
Lucas	5107010	52.0	60.0
Siemens	2-007-63656	55.0	63.5
MSD	2014	72.0	83.1
GM/Bendix (red)	25500139	82.0	94.7
MSD	2015	96.0	110.9

are not unheard of depending upon the manufacturer. Remember, most fuel injectors are rated at 3-bar (43.51 PSI), while the LS1 fuel system operates at 4-bar (58.02 PSI). This is important to remember since you will need to convert the injectors' advertised flow rate at 3-bar to the LS1's 4-bar operating pressure for proper sizing in the LS1.

Pulse Width

The PCM controls the quantity of fuel supplied to the engine by regulating the amount of time the fuel injector is open. This time is known as pulse width and is measured in milliseconds. At wide-open throttle, the fuel system has a specific amount of time in order to deliver the proper amount of fuel at any given rpm. As the rpm goes up, that window of opportunity for fuel delivery grows smaller. For example, at 4000 rpm, there is 30ms to deliver sufficient fuel, but at 6000 rpm, there is only 20 ms available before an injector is no longer being cycled but held open. When this happens, an injector is said to be at 100% duty cycle, or "static." Pulse width is calculated with the following equation:

Pulse Width = (60,000 / rpm) x 2

Duty Cycle

Injector duty cycle is simply the measure of actual pulse width to maximum pulse width. Simply, this means that an injector with a 50% duty cycle (for a given RPM) is open 50% of the time. An injector held open (static) is said to be at 100% duty cycle or has reached it's maximum pulse width for given RPM. Under most circumstances, duty cycle should not exceed 80-85% (due largely to injector solenoid heat buildup), although some are able to run at 100% for short periods of time. However, doing so for an extended period will increase the likelihood of damaging the injectors by overheating them. This can cause irregular fuel delivery or a erratic operation at low engine rpm. Irregular fuel injector operation not only makes for poor performance, but also makes it difficult to properly tune your engine for maximum performance.

Brake Specific Fuel Consumption

Brake Specific Fuel Consumption (BSFC) is a variable that estimates an engine's efficiency at converting fuel to horsepower. BSFC is the ratio of fuel an engine consumes in pounds per hour for each horsepower produced. The number should be from 0.4 to 0.6 lbs.-hr. for gasoline-powered engines. A BSFC of 0.5 lbs.-hr. is a reasonable and conservative initial estimate for most naturally aspirated engines.

Engines that are turbocharged, supercharged or nitrous-injected should run a richer air/fuel ratio, which means the BSFC estimate should be increased; .60 is a commonly accepted starting point. Accordingly, a forced induction engine will require larger injectors to generate the same horsepower as a naturally aspirated mill. Naturally aspirated race engines can be run as lean as .40, but this requires the tune to be perfect. However, leaner doesn't always mean more performance.

The following are some general guidelines when choosing a BSFC number:

- Nitrous-injected, turbo and supercharged engines: 0.60
- Naturally aspirated street engines: 0.50
- Naturally aspirated performance engines: 0.45
- Naturally aspirated race engines: 0.40

Fuel Injector Selection

Choosing fuel system components can be a confusing and frustrating experience for many people, but it need not be a hassle. By systematically addressing the following issues, you can determine the ideal fuel system for your engine. Before you start, you need to know a few things about your ultimate goal, including desired power level, vehicle weight and performance goals. For the purposes of this exercise, let's say you have a 3500-lb. car that you want to run 10.50's at about 130 mph, without the use of a power adder.

The Lucas "-009" injector is an excellent performer that is rated to flow 42.5 lbs.-hr. at 3-bar, and 49.1 lbs.-hr. at 4-bar. In the real world, that is enough to support well over 600 horsepower.

Step 1: Determine Power Goal

There are several formulas available to calculate horsepower based on ET or trap speed and vehicle weight; I will present two of the best. Method one tends to be slightly less accurate because elapsed time depends heavily on traction and how well the car leaves. Therefore, method two is the preferred equation.

Method 1: Calculate Horsepower From ET and Weight
Weight / ((Elapsed Time / 5.825) ^3) = HP
Example: 3500 / ((10.50 / 5.825) ^3 = 597.57 HP required

Method 2: Calculate Horsepower From Trap Speed and Weight
((Trap Speed [mph] / 234) ^3) X Weight = HP
Example: ((130 / 234) ^3) X 3500 = 600.14 HP required

These formulas tells us that we need approximately 600 horsepower at the flywheel to run the numbers.

Step 2: Determine Fuel Requirements

We now know that it will take about 600 HP to propel our 3500 pound vehicle to 10.50's at 130 mph. The fuel requirement formula allows us to determine how much fuel is required to make our 600 HP. Remember that engines with power adders should use .6 as the BSFC multiplier.

Horsepower x BSFC = Fuel Required

Example: 600 HP x .5 lbs.-hr. = 300 lbs.-hr. of fuel required

Step 3: Determine Minimum Injector Capacity

Now that we know how much fuel the engine requires, we can determine the flow capacity needed. We do this by dividing fuel volume by the number of injectors. In this case, eight 37.5 lbs.-hr. injectors are the minimum necessary to support our 600 hp engine (at 100% duty cycle).

Fuel Required (lbs.-hr.) / number of injectors = Injector Capacity

Maximum Horsepower at 4-bar (100 / 80% Duty Cycle)

Injector	BSFC 0.4	BSFC 0.5	BSFC 0.6
1997-98 LS1	572/458	458/366	382/305
1999-2000 LS1	536/429	429/343	357/286
2001-03 LS1/LS6	572/458	458/366	382/305
Bosch 30	692/554	554/443	461/369
Bosch 36	832/666	666/532	555/444
Lucas 42.5	982/786	786/629	655/524
MSD 50	1154/923	923/739	769/615
Siemens 55	1270/1016	1016/813	847/677
MSD 72	1662/1330	1330/1064	1108/886
Bendix 82	1894/1515	1515/1212	1263/1010
MSD 96	2218/1774	1774/1420	1479/1183

Note: Data provided reflects approximate horsepower levels at 100%/80% duty cycles.

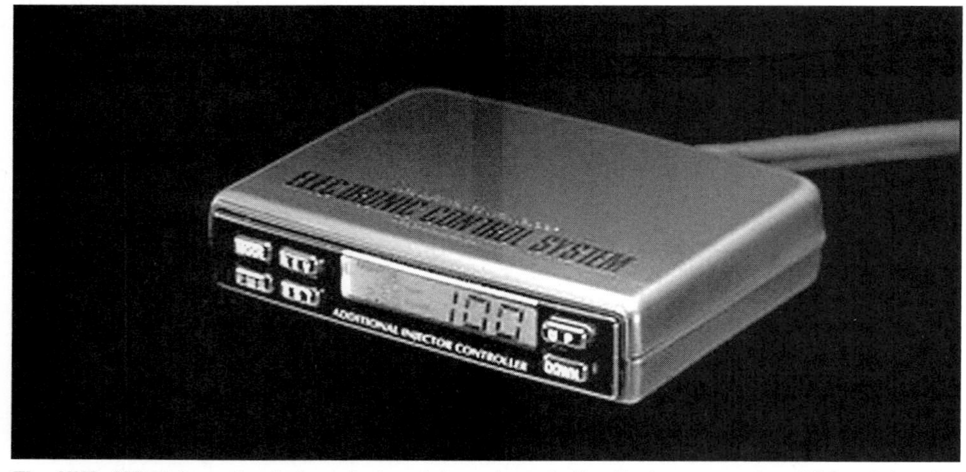

The HKS AIC-III is a stand-alone fuel enrichment controller that can supply additional fuel for forced induction engines.

Example: 300 lbs.-hr./8 = 37.5 lbs.-hr. injector capacity

Step 4: Determine Injector Size

It's unrealistic to establish the fuel flow to the engine based on an injector duty cycle of 100% (wide open all the time), as this virtually guarantees reliability issues. This formula uses injector duty cycle based on 80%. The 0.8 multiplier of the Injector Size formula helps us derive a practical, safe maximum Injector Flow

Injector Capacity / Duty Cycle = Injector Size

Example: 37.5 lbs.-hr. / .80 DC = 46.87 lbs.-hr. injectors

Step 5: Select an Injector

We have determined that it takes 300 lbs.-hr. of fuel to generate 600 HP, and we know that 47 lbs.-hr. injectors are the minimum required to deliver that amount at 80% duty cycle. In many cases the nearest injector size will barely meet the performance requirement. When choosing injectors, be sure to always round up to the next largest size.

The chart above shows the maximum horsepower various inject-ors will support based on their flow rate at 58.02 psi (4-bar) and various BSFC values. The injectors are listed in the table by their 3-bar ratings, because this is the industry standard. Maximum Horsepower Potential is calculated like this:

At 100% Duty Cycle = (injector flow rate X 8) / BSFC
At 80% Duty Cycle = (injector flow rate X 8 X .8) / BSFC

Supplemental Fuel Systems

Several manufacturers offer add-on fueling setups, sometimes called 9th injector systems. These items are not used as often as they were in the past due to the wide assortment of injectors and fuel management systems now available. These systems are still a good option, however, as they offer good horsepower capability and are easy to tune.

Adjustable Fuel Pressure

In the past, the easiest way to increase injector capacity was to raise the fuel pressure, but this is not possible with the LS1s without an overhaul of the returnless fuel system (1997 Corvette excepted). The advantage of converting to an adjustable fuel pressure system is the

The OEM fuel rail is fabricated from stainless steel, and can be converted to a traditional vacuum-modulated system fairly easily.

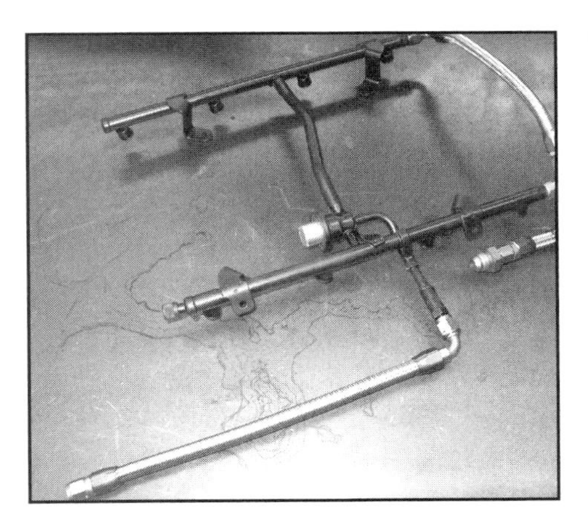

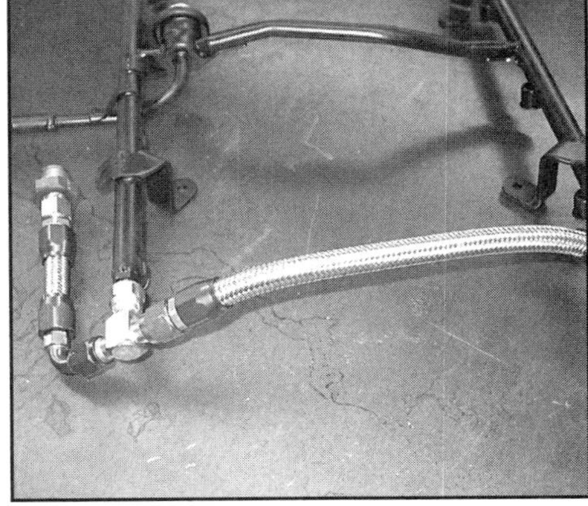

A stock fuel rail can be converted to utilize an adjustable fuel pressure regulator.

A converted system will require an adjustable fuel pressure regulator like this one from Aeromotive.

ability to easily tailor fuel pressure to meet your combination's needs. If you do utilize an adjustable fuel system, an easy rule of thumb is that a 1 psi increase in fuel pressure will increase injector flow rate approximately 1%. Remember that raising the fuel pressure will richen the entire rpm range from idle to rev-limiter, which may lead to less than optimum air/fuel ratios when not at WOT.

There are a couple of options if you choose to make this modification. The first and simplest option is to use the fuel rail from a '97 Corvette and add an adjustable fuel pressure regulator. This will also require the addition of a fuel return line from the rail back to the tank, and the bypass of the stock regulator at the fuel tank.

You can also convert your existing fuel rail to a traditional return-style system, or you can use aftermarket universal fuel rails. Another option is the aftermarket fuel rails for a 5-liter Mustang, which use the same injector spacing as the Gen-III, but must be mounted "backward" and require fabrication of mounting brackets. All of these options will require the purchase of an adjustable fuel pressure regulator, modification of the internal regulator in (or replacement of) the in-tank pump, and installation of a return line from the fuel pressure regulator to the tank.

Boost-indexed fuel pressure regulators utilize a boost/vacuum signal from the manifold to act upon an internal diaphragm that changes fuel pressure as manifold pressure changes at a 1:1 ratio to boost. In other words, if the base fuel pressure is 58 psi, and the manifold boost is 10 psi, the total fuel pressure should then increase to 68 psi.

Fuel Pumps

There are two broad categories of fuel pumps available for EFI engines: in-tank and in-line. In-tank pumps simply replace the factory pump on the hanger assembly inside the fuel tank. These are quiet and reliable, but have their limitations when extraordinary amounts of fuel are required.

Inline pumps are external, and are sometimes used in conjunction with an in-tank unit. This method is widely

A high-end in-line pump like this Aeromotive Pro-Series will flow more than 900 lbs.-hr. at 13.5 volts and 45 PSI. That's enough fuel to support up to 1800 horsepower.

A single Walbro 340 in-tank pump like this will support up to 500 rear wheel horsepower. It's all the pump most cars will ever need.

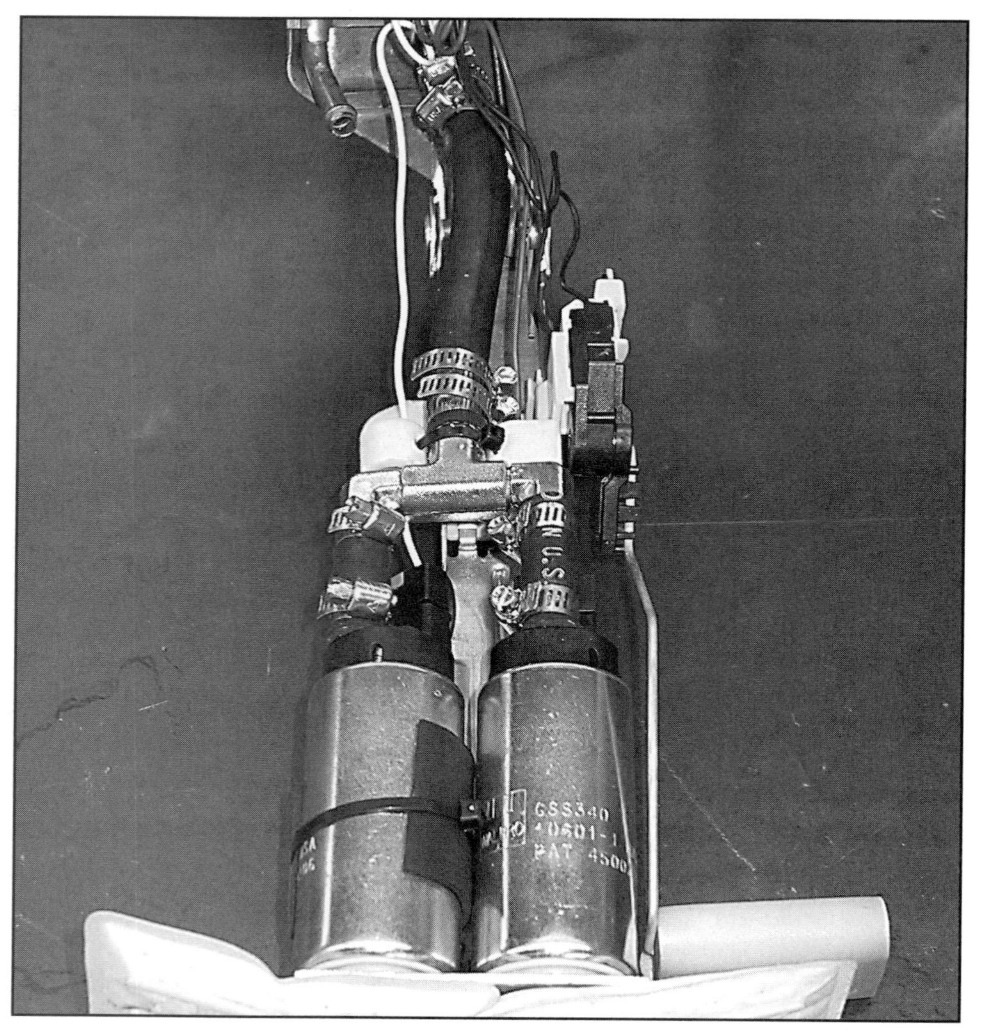

For those that need more, a dual in-tank system is a great option.

used and it does take much of the burden off of the in-tank pump. However, maximum fuel delivery is limited by the weakest pump, which, in most cases, is the intank pump. In-line pumps offer the ultimate in terms of fuel delivery, but tend to be noisy and require additional plumbing, so if stealth is what you're after, look at the intank pumps.

Dual In-Tank Pumps

Dual in-tank pumps are an excellent alternative to in-line pumps, as they place in-tank pumps on a single factory pump hanger assembly. Dual pumps can be run constant, or staged through the use of an RPM-activated switch or a Hobbs switch. Dual in-tank pumps resolve most of the limiting issues of a single in-tank pump, and require no additional plumbing.

The stock fuel line and filter is sufficient for a single intank fuel pump.

High octane racing fuel does not contain more energy than pump gas. It does provide a slower burn rate, delaying the onset of detonation in high compression or power adder engines.

Boost-a-Pump

Another method used to extract a little more capacity from an existing fuel system is to use a Boost-a-Pump product. This electronic product simply increases voltage to the fuel pump to "over-drive" it a bit, increasing its fuel delivery capacity. In many cases, these units are boost referenced for added volume delivery as needed.

Determining Fuel Pump Requirements

Because flow decreases as pressure increases, you should verify the fuel pump you choose will flow the required volume at 58 PSI (or whatever pressure you intend to run).

In Step 2 above, we calculated the approximate fuel needs for our 600 horsepower engine based on this equation:

Horsepower x BSFC = Pounds per hour
Example: 600 hp x .5 = 300 lbs.-hr.

Determine gallons per hour by dividing pounds per hour by the fuel mass. Gasoline weighs 5.8 to 6.5 pounds per gallon, so we will use an average of 6.15 pounds per gallon for this exercise.

Pounds per hour / gasoline weight = Gallons per hour
Example: 300 lbs.-hr. / 6.15 lbs. = 48.8 GPH

Once you have determined the amount of fuel required, the next step is to choose the right fuel pump. Fuel pump manufacturers flow rate their pumps at all different pressures and voltages, so comparing them can be tricky. Do your homework to be sure the pumps you compare are rated under the same conditions so you compare apples to apples.

Fuel Line

The stock fuel line is sufficient for all but the most radical of combinations. However, if you elect to convert your fuel system to one utilizing an adjustable fuel pressure regulator, you might find it an opportune time to upsize the supply line to AN –8 braided stainless steel line and utilize the OEM supply line as the return.

Water/Alcohol Injection

A forced induction engine often has an appetite for high-octane (and high dollar!) race fuel. Injecting alcohol into the intake air stream of a forced induction engine has a substantial effect similar to that of an intercooler. Cylinder temperatures can drop as much as 300 degrees, which reduces detonation and allows higher boost levels to be run with lower octane fuel. Tuning a water/alky system is relatively simple, but you need to be sure not to run the reservoir dry, or you risk severe detonation and serious engine damage.

10

Ignition and PCM Tuning

If you're a fan of the LS1, you should also appreciate the incredible job its computer does in dosing out power. As the vehicle computer systems, control sensors, and onboard diagnostics have become more sophisticated and powerful, they have also in turn become more interdependent and complex. Initially, the sophistication of the Powertrain Control Module (PCM) made it difficult for the performance aftermarket to "crack" the PCM to make the tuning alterations necessary to support significant changes for higher outputs.

Like most engine/driveline management systems of the last decade, the GM PCM controls a wide variety of vehicle operating parameters. And it does it all flawlessly, without grumbles, rumbles or hesitation over an endless variety of conditions. The operating parameters and range provided by GM in the LS1's PCM has dealt admirably with minor performance changes such as high flow intake and exhaust systems. The availability of aftermarket-calibrated MAF meters and the MAF Translator has a long way to go in tricking the PCM into providing altered fuel and ignition timing characteristics required to deal with modifications such as larger fuel injectors or performance camshafts.

In many cases however, the power gains seen upon initial installation and testing on the dyno have disappeared after several hundred miles of driving as the PCM has proven itself capable of relearning, in order to again achieve its target figures as calibrated by GM. And when radical changes are made, the PCM will figuratively throw its hands in the air, illuminating the dreaded Check Engine light. Thankfully, the aftermarket now offers several ways for enthusiasts to tame the Gen-III PCM.

PCM

The Powertrain Control Module (PCM) system for the LS1 is the brain for one of the most expansive serial communication networks deployed on any GM vehicle to date. The network is comprised of a dedicated communication system tying together up to eight other systems within the vehicle. These systems include HVAC, instrument panel, Anti-Lock Braking

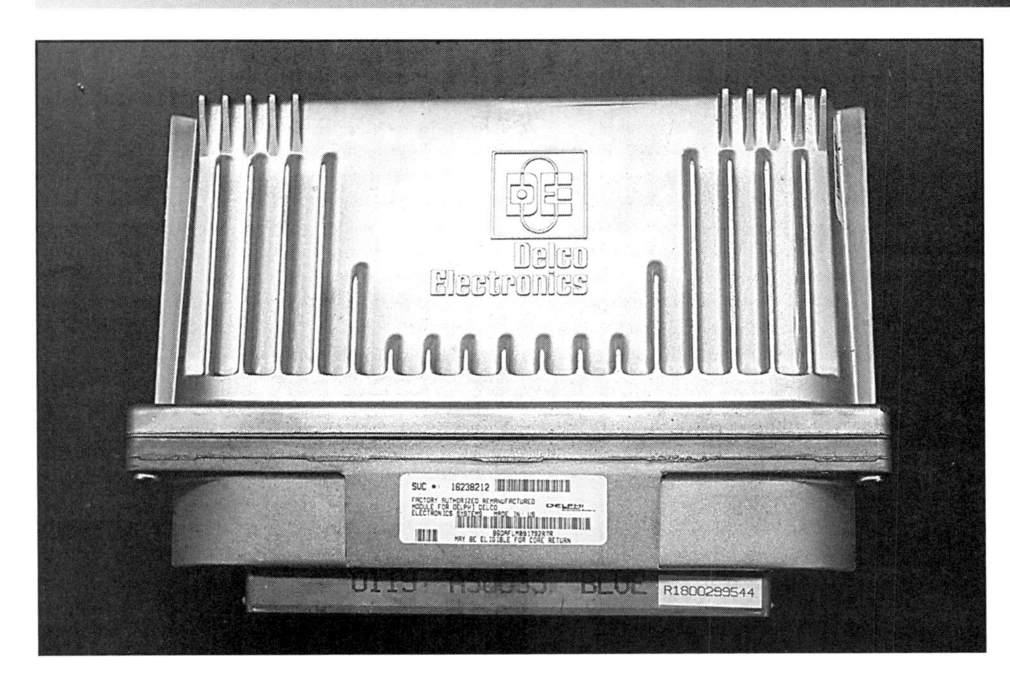

The Gen-III PCM, initially at least, was very difficult to tune. Times have changed, and the PCM has proven to be incredibly flexible.

System (ABS), the Electronic Throttle Control (ETC) system (in Corvettes) and other vehicle subsystems. Another feature of this control system is a standard platform-to-power train interface that reduces the complexity of utilizing this controller in other vehicles. This is achieved by standardized inputs and outputs to have common interface definitions.

The PCM incorporates a 32-bit processor and a digital signal-to-noise enhancement filter. This converts the calibration for the knock system from a separate plugable hardware circuit to calibration values within the computer code. This method has greatly enhanced the performance of the knock system by permitting dynamic tuning of the filter so that noise can be precisely filtered out, thus allowing optimal spark timing under all conditions. Because of the LS1's unique (at the time of introduction) dedicated ignition coil per cylinder configuration, no existing PCM within GM was capable of controlling this engine. Software development had to have basic functionality complete in order for the first engine to be operational. For this to happen, initial software development relied heavily on dynamic vehicle and engine simulations. Once engines became available, development continued using both the actual engines and simulation tools.

Software was developed using a combination of Modula-GM and M68030 assembly languages and structured design methodologies. A number of math-based models make up a portion of the software. These include prediction of airflow, real-time dampening for active suspension control, wall wetting to predict transient fueling requirements and a torque model that predicts the engine torque output. The final product contains 375K of code, 75K of which is calibration code located in ROM, and the processor uses 13.5K of RAM. It also contains 512K of data for the PCM run-time code for LS1s from 1997 to 2002. Beginning with 2003, this has grown to 1024K of data. All these data tables are stored on what is called a Flash EPROM chip.

SFI

Sequential-Port Fuel Injection (SFI) is the state-of-the-art method to control an electronically fuel-injected engine. SFI uses one fuel injector per cylinder for optimum fuel distribution. Unlike batch-fire systems where multiple injectors are fired simultaneously once per engine revolution, SFI fires each injector individually and sequentially time the cylinder's intake cycle. SFI maximizes power production, optimizes fuel economy, and has the flexibility to meet ever-tightening emissions-control requirements.

The PCM controls all aspects of engine operation, most importantly fueling and ignition timing. It sets the fuel delivery and ignition schedule by applying data, such as crankshaft position, mass airflow and intake air temperature, engine speed, coolant temperature and many other parameters, to fuel look-up tables in the PCM calibration. Based on those look-up tables, each cylinder's injector is pulsed in the engine's firing sequence such that a precisely metered amount of fuel is delivered to the intake port just before the valve opens. The injector design and location produce a spray pattern that contributes to the engine's smooth idle and fuel efficiency.

Under most driving conditions, the PCM operates under a closed loop feedback mode to trim fuel delivery to optimum air/fuel ratio levels. Free oxygen in the exhaust is an accurate indicator of fuel mixture and is measured by four oxygen sensors in the exhaust manifolds; two upstream

The dreaded Check Engine light means that OBD-II has detected something gone awry.

of the catalytic converters, and two downstream. The O2 sensors relay oxygen content information to the PCM, which then adjusts mixture to optimize engine performance. The Mass Airflow Meter (MAF) is essential for accurate fuel delivery. The MAF constantly measures the engine's air intake under varying conditions, such as changes in load, altitude and temperature and signals the PCM to make the necessary adjustments.

OBD-II

The LS1 PCM has more computing power than previous units, which allowed the new engine to meet more stringent emissions regulations in, the late 90s and early 00s. The PCM is equipped with second-generation On-Board Diagnostics (OBD-II). The most important feature of OBD-II is that it requires the PCM to predict potential failures of emissions controls as well as immediate notification of failures that have already occurred. Two significant features enabling this prediction are catalyst monitoring and misfire detection.

Misfire detection demands a sophisticated processor in the PCM, very accurate crankshaft position data and some technically innovative software. The PCM reads minute variations in crankshaft speed as the engine accelerates and decelerates in reaction to power impulses. Inconsistent variations in these rates are indicative of engine misfire. Specific types and durations of misfire can be a sign of other emissions system problems and will turn on the Malfunction Indicator (MIL) or Check Engine lamp. The trick comes in accurately determining what is a misfire and what isn't. In low-level engine misfire, things like bad fuel, cracked spark plug insulation, plugged fuel injectors, or even a mild, lopey, performance cam can all lead to a misfire diagnosis that's picked up by OBD-II.

OBD-II can mean easier, less expensive repairs for consumers. If a problem is detected, a Diagnostic Trouble Code (DTC) is stored and indicates the type of fault detected. The PCM stores and retrieves diagnostic messages to help technicians fix problems quickly and accurately. The PCM alerts the driver by illuminating the Service Engine Soon light when it detects a problem in any of the monitored components. Unlike first-generation systems that signaled only system failures, OBD II alerts the driver to have the vehicle serviced before experiencing a possible breakdown or incurring more expensive repairs.

Sensors

In order to provide the correct amount of fuel for every operating condition, the PCM has to monitor many input sensors. The PCM uses a dual sensor method to establish engine position. One sensor is used to read a dual track encoded target wheel mounted to the crankshaft. The crank sensor and target wheel mounting location is located in the block near the quietest deflection point on the crankshaft, thus providing the lowest dynamic air gap variation possible. It is located above the starter on the passenger side, at the rear of the block. Encoding the target wheel permits quick identification of individual cylinders and accurate location for precision of the spark and fuel delivery. The second sensor is the cam sensor, which is located on the top of the block behind the intake manifold. The cam sensor is utilized to determine which half of the engine's firing sequence is currently in operation. Both sensors use magneto-resistive sensing technology, which permits very accurate position sensing down to zero crank and cam speed.

Detonation

Detonation protection is similar to what's been used in the past. However, the amount of sensors and location has changed from previous small-block Chevy implementations. There are now two knock sensors working in a feedback system with the PCM. When a knock sensor "hears" detonation, the PCM retards timing and then waits for additional sensor input. If the detonation stops; timing is gradually reset to the value called for in the PCM calibration tables. If detonation continues, timing is retarded an additional amount. These dual knock sensors are now mounted under the intake in the engine valley for better response and isolation from external noise. Previously, a single knock sensor was

The PCM Sensors

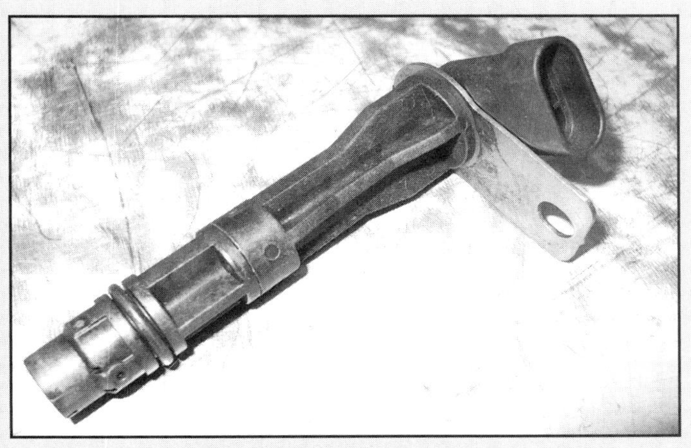

Cam Position Sensor: Tracks position of camshaft, and is used to detect misfires. It is located at the rear of the block, behind the intake manifold and above the camshaft.

Coolant Temperature Sensor: This sensor allows the PCM to determine when the engine has reached or is exceeding normal operating temperature. This sensor has a wide range of operating functions, from telling the PCM when to use startup enrichment mode to telling the PCM when to turn on the radiator cooling fans. It also supplies data to operate the dashboard coolant temperature gauge. The coolant temperature sensor is located on the driver's side cylinder head.

Crank Sensor: It monitors engine speed, which is one of the factors used to calculate injector pulse width and ignition timing. It is located above the starter, at the rear of the block on the passenger side.

Idle Air Control Valve: Not a sensor, but the PCM controls idle rpm with the IAC valve. To increase idle rpm the PCM moves the IAC valve out allowing more air to pass by the throttle body. To decrease rpm it moves the IAC valve in to reduce air past the throttle body. The IAC is located on the throttle body.

Intake Air Temperature Sensor: The PCM uses this information to alter timing and fueling curves. It is located in the intake tract, usually on the airbox or integral to the mass airflow sensor.

Sensors, *continued*

Knock Sensors: Designed to detect the high-frequency vibrations caused by detonation. By employing two knock sensors, the PCM can operate the engine very near its detonation limit; thereby improving power and efficiency. The knock sensors are located beneath the intake manifold, on the engine valley cover.

Manifold Absolute Pressure Sensor: Monitors the pressure of the air in the intake manifold. The amount of air being drawn into the engine is a good indication of how much power it is producing; and the more air that goes into the engine, the lower the manifold pressure, so this reading is used to gauge how much power is being produced. It is located at the rear of the intake manifold.

Mass Airflow Sensor: Information sensor that measures the quantity of air flowing into the engine. This allows the engine management system to accurately calculate the air/fuel ratio and optimize engine performance. The MAF is situated between the airbox and the throttle body.

Oil Pressure Sensor: Monitors engine oil pressure and will illuminate the Check Engine light or kill the engine if oil pressure drops to zero. It also supplies information to operate the dashboard oil pressure gauge. It is located behind the intake manifold at the rear of the block, next to the cam position sensor. There is also an Oil Temperature Sensor (not shown) that monitors oil temperature in Corvettes and allows the PCM to reduce power if maximum temperature is exceeded. The oil temperature sensor is located in the Corvette's oil pan.

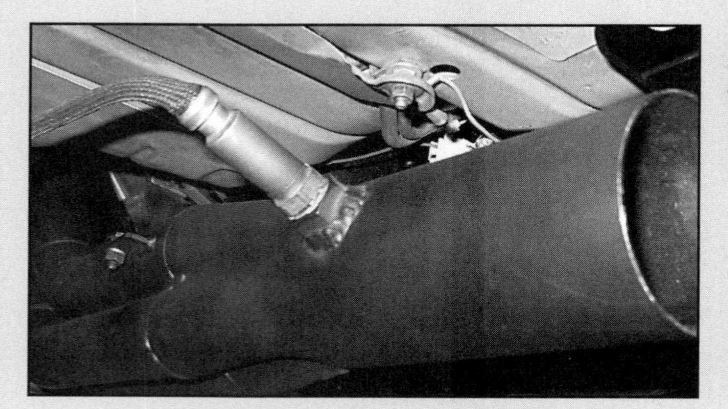

Oxygen Sensors: These monitor the amount of oxygen in the exhaust so the PCM can determine how rich or lean the fuel mixture is and make adjustments accordingly. Comparing data from the four oxygen sensors screwed into the exhaust system; two upstream of the catalytic converters, and two downstream, also gives indication of catalytic converter performance.

Throttle Position Sensor: Monitors the throttle valve position so the PCM can respond quickly to changes, increasing or decreasing the fuel rate as necessary. Corvettes have their TPS mounted at the throttle pedal; others are on the throttle body. The Vehicle Speed Sensor monitors vehicle's road speed for cruise control, fuel trim, and speedometer operation. The VSS is located inside the transmission.

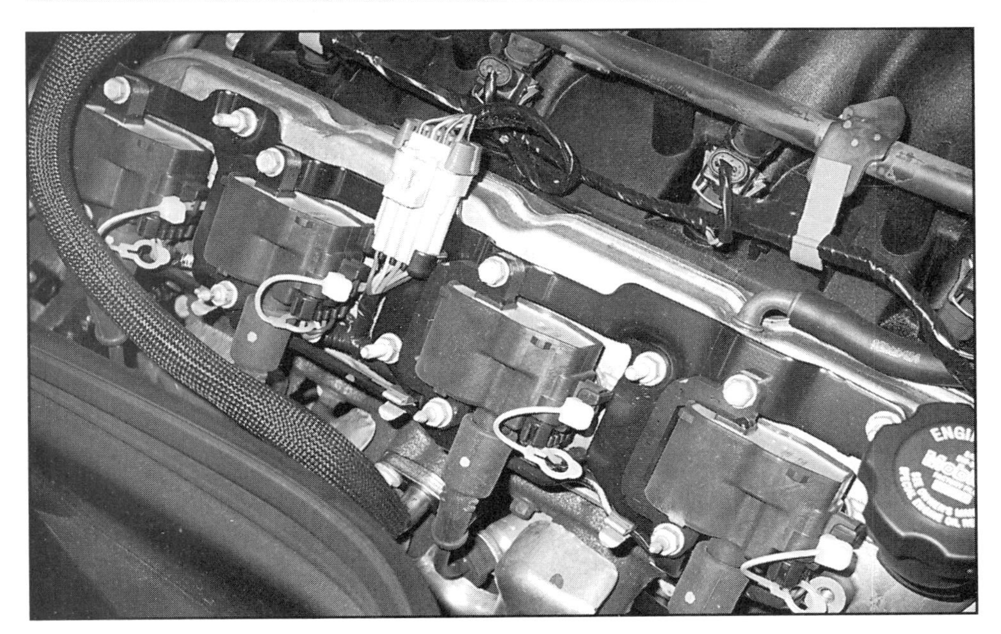

Gen-III's use one coil pack per cylinder. Aftermarket coils are beginning to appear on the market, but there is no evidence that a performance gain will be had by making this change.

GM's Tech-II service tool is the ultimate diagnostic piece for any GM vehicle. Though all GM service departments use them, the only way for an enthusiast to purchase one is on the gray market. In most cases, it is a prohibitively expensive tool for the average enthusiast.

mounted on one side of the block.

Ignition

Mounted on the valve covers of the Gen-III engines, the cylinders have dedicated coil and coil driver assemblies with short plug wires connecting each spark plug. The reasons for moving the coils to the valve covers are because less spark energy is dissipated by short spark plug wires, so more energy is available at the plug. In fact, this increases energy delivery to nearly 100%. The shorter plug wires also reduce radio frequency interference with onboard computers and the sound system. The drive electronics are located in the coil, which removes the need to dissipate heat energy within the PCM or add additional interface wires to the engine package for remote mounted drive electronics.

The rest of the ignition system is conventional. The PCM controls ignition advance by computing the optimum trigger point and sending a trigger impulse to the coil driver at the appropriate cylinder. Which brings us to the other notable ignition feature, a new firing order: 1-8-7-2-6-5-4-3. This lessens vibration and stabilizes the idle. Standard spark plugs used in the LS1 are AC-Delco platinum-tipped units.

Scan Tools

Scan tools should be a component in every EFI hot-rodder's toolbox. A scan tool plugs in to the car's diagnostic port beneath the dashboard and allows the user to observe the PCM's activity in running the engine as well as retrieve diagnostic codes that may have been set and stored.

Tech-II

The most exotic scan tool of them all belongs to General Motors. It is known as the TECH-II service tool. The TECH-II cannot only monitor all vehicle systems, but also can perform diagnostic tests and even upload calibration revisions. This is one powerful tool that everyone would love to have. Unfortunately, GM does not offer them for sale to the public.

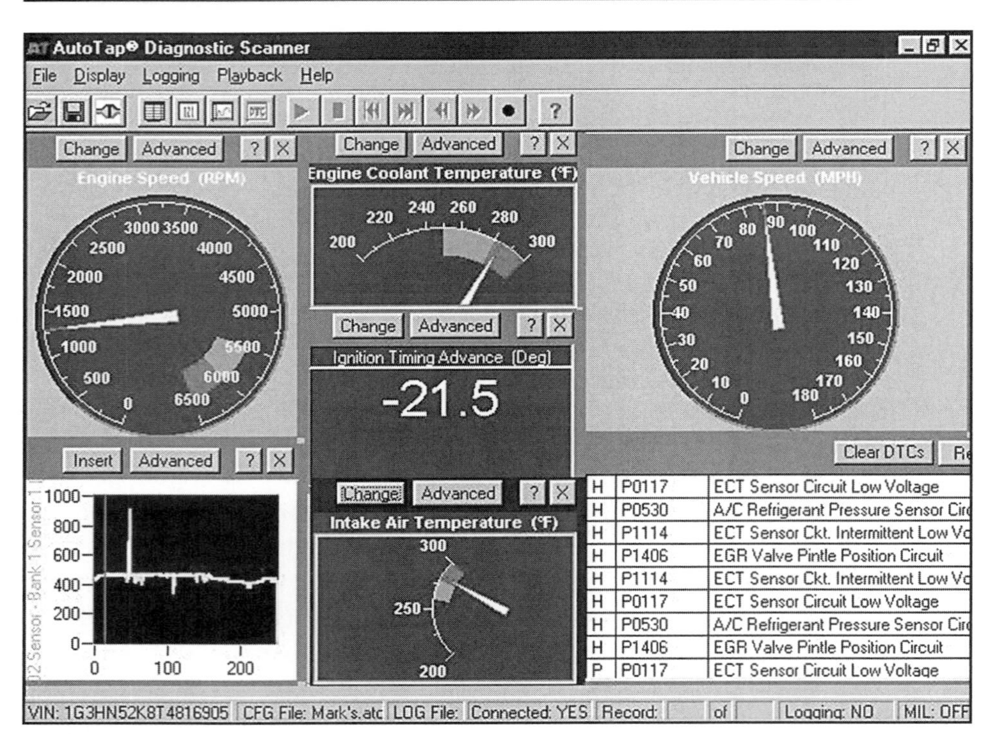

Autotap software works equally well for tracking down OBD-II error codes as it does for engine performance data logging.

Autotap

Autotap is PC-based software that offers real-time display and recording capability from all of the vehicle's sensors. It allows you to monitor such things as spark advance or retard, injector pulse width, and MAF rate, as well as hundreds of other real time parameters from the PCM. It also allows you to read the diagnostic trouble codes and reset them once a repair has been made. Autotap is the quintessential scan tool for LS1 performance.

EFILiveV5

EFILive is another PC-based software package that uses the same electronic interface module as Autotap, but was designed with the LS1 in mind. Its Pro version has some real-time bi-directional controls that can further aid in the tuning of your LS1. It functions much like Autotap, but with a completely different visual interface and LS1 specific guides.

Both of these tools are considered OBD-II diagnostic tools. Autotap's added benefit lies in that it can interface with any GM OBD-II capable vehicle where as EFI-LiveV5's benefit lies in its bi-directional controls.

PCM Calibration

PCM calibration is one of those things that send many would-be EFI hot-rodders into fits of nervous convulsions. Tip-in stumble, poor idle quality, and difficult cold starts can be daunting, no doubt. Add expense and long lead times into the equation, and it's no wonder most people take a pass.

It should come as no surprise that the algorithms used by the PCM to control the engine are quite complicated. The software has to allow the car to satisfy emissions requirements for 100,000 miles, meet EPA fuel economy requirements and protect engines against abuse. The PCM uses algorithms with hundreds of lookup tables to determine the injector pulse width for given operating conditions. The equation will be a series of many factors multiplied by each other. Many of these factors will come from lookup tables. For example, this equation will only have three factors, whereas a real control system might have a hundred or more.

Pulse Width = (Base Pulse Width (RPM, Load)) x (Factor A) x (Factor B)

Calculating Pulse Width

In order to calculate the pulse width, the PCM first looks up the base pulse width in a lookup table. This is a function of engine speed (rpm) and load (which can be calculated from manifold absolute pressure). Let's say the engine speed is 2000 rpm and load is 4. We find the number at the intersection of 2000 and 4, which is 8 milliseconds.

So, since we know that base pulse width is a function of load and rpm, and that pulse width = (base pulse width) x (factor A) x (factor B), the overall pulse width in our example equals:

Example: 8 x 0.8 x 1.0 = 6.4 milliseconds

From this example, you can see how the control system makes adjustments. With parameter B as the level of oxygen in the exhaust, the lookup table for B is the point at which there is (according to engine designers) too much oxygen in the exhaust; and accordingly, the PCM cuts back on the fuel.

A high-resolution wideband oxygen sensor is commonly used as an air/fuel ratio-monitoring device when tuning on a dynamometer.

Hypertech's Power Programmer was the first handheld available for the Gen-III. Its use is best kept limited to gear and fan calibration and other minor changes.

Real control systems obviously have many more parameters, each with their own look-up tables. Some of the parameters even change over time in order to compensate for changes in the performance of engine components like the catalytic converters or to adjust to changing weather conditions. And depending on the engine speed, the PCM may have to do these calculations over a hundred times per second.

Generally, the more aggressive the combination (cam, head flow, valve size, displacement) the more critical the PCM tune. While there are those that will tell you it is unnecessary with a heads and cam package, it really depends upon how far you plan to take your combination. Let's face it, the best way to maximize the performance of any modified fuel injected engine is through custom PCM calibration.

Air/Fuel Ratio

When major combination changes are made, a change to the PCM calibration is often necessary. A perfect air/fuel ratio is what differentiates an engine with sharp, crisp throttle response, from a stumbling, erratic mess. Some of the others are proper ignition timing and intelligent engine component selection. It is the quest of a perfect air/fuel ratio and timing for any given rpm that leads us to our discussion of custom PCM tuning.

The flash memory area of the PCM holds all of the lookup tables. The tables in an aftermarket tune contain values that result in modified fuel rates and altered ignition curves during certain driving conditions. For instance, they may supply more fuel at full throttle and at every engine speed. They may also change the spark timing, perhaps only at part-throttle. Since the performance tuners are not as concerned with issues like reliability, mileage and emissions controls as GM, they use more aggressive values in the fuel and ignition maps of their performance calibrations.

Handheld Programmers

Handhelds work well for adjusting

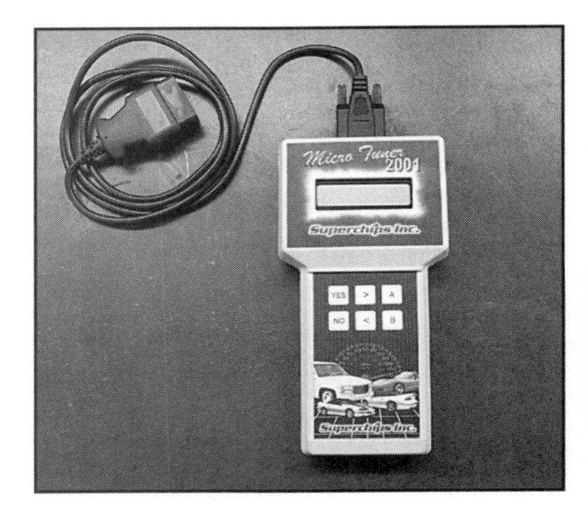

The Superchips Micro-Tuner handheld programmer lets the user make simple adjustments or upload a custom calibration.

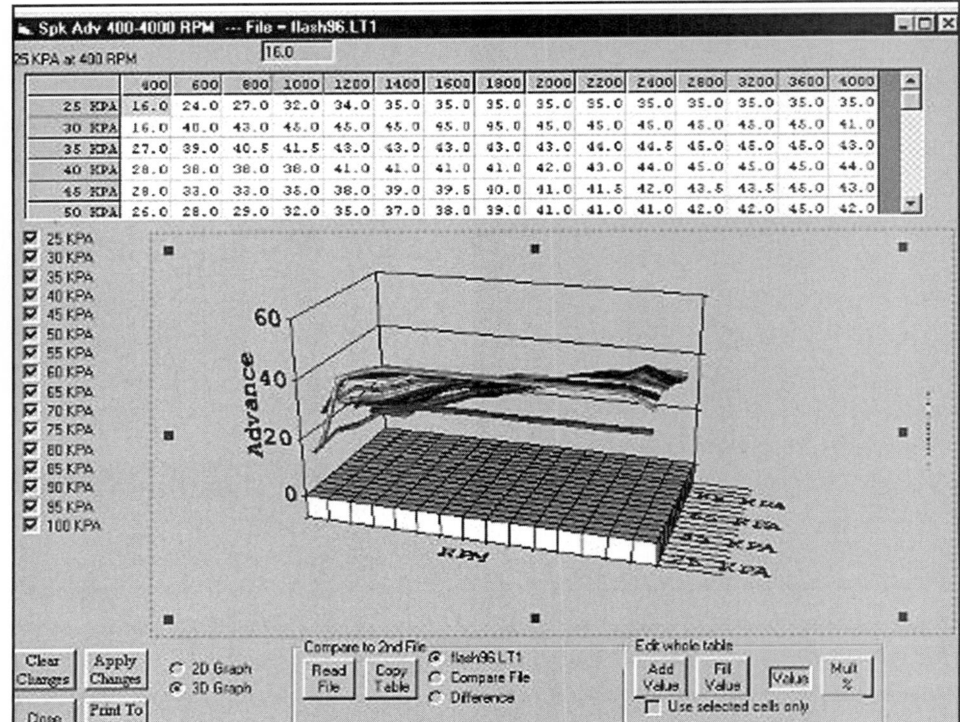

	400	600	800	1000	1200	1400	1600	1800	2000	2200	2400	2800	3200	3600	4000
25 KPA	16.0	24.0	27.0	32.0	34.0	35.0	35.0	35.0	35.0	35.0	35.0	35.0	35.0	35.0	35.0
30 KPA	16.0	40.0	43.0	45.0	45.0	45.0	45.0	45.0	45.0	45.0	45.0	45.0	45.0	45.0	41.0
35 KPA	27.0	39.0	40.5	41.5	43.0	43.0	43.0	43.0	43.0	44.0	44.5	45.0	45.0	45.0	43.0
40 KPA	28.0	38.0	38.0	38.0	41.0	41.0	41.0	41.0	42.0	43.0	44.0	45.0	45.0	45.0	44.0
45 KPA	28.0	33.0	33.0	35.0	38.0	39.0	39.5	40.0	41.0	41.5	42.0	43.5	43.5	45.0	43.0
50 KPA	26.0	28.0	29.0	32.0	35.0	37.0	38.0	39.0	41.0	41.0	41.0	42.0	42.0	45.0	42.0

LS1 Edit is the tuning method of choice among hardcore LS1 performance enthusiasts.

the speedometer for rearend gears, and modifying the temperature at which the cooling fan is activated. While most offer an off the shelf performance tune program, these typically limited performance gains and it is not unheard of for these programs to actually decrease power under some conditions. Because they are generic programs that usually have a bit more timing without much extra fueling, the engine runs closer to (and oftentimes over) the detonation threshold.

Custom Tuning

The first alternative to the handheld programmers is to send your PCM to a custom calibrator like Fastchip, The Turbo Shop or Z-Industries. These companies will take information about such things as cylinder head flow numbers, fuel injector size, operating rpm, gears, converter stall speed, and car weight to devise a calibration to control engine combination seamlessly. The downside is that the PCM must be removed from the car and sent to the tuner. Obviously this is a problem if your car is a daily driver, as you must wait for them to "flash" and return it. You could get a spare PCM from a junkyard to minimize downtime.

PCM calibrators can do basic things such as disable SES lights caused by things such as EGR removal, delete the Skip Shift "feature," and far more complicated procedures such as completely re-map the fuel and ignition curve look-up tables. Other things a custom PCM tuner can alter:

- Air /Fuel Ratio table
- Base Spark Advance
- Diagnostic tests modes
- Fan Temperature
- Fuel Injector flow rate Manifold vacuum
- Fuel injector Offset table
- Fuel Pressure versus voltage
- Gear and tire size scaling
- Idle Speed tables
- MAF calibration table
- Power Enrichment vs. Temperature table
- Power Enrichment versus RPM table
- Rpm Limiters
- Spark Advance vs. rpm vs. Load
- Knock Retard tables
- Torque Management tables
- Vehicle Anti-Theft System disable
- Volumetric Efficiency table
- Wide Open Throttle Hot & Cold tables

Fuel and Ignition

Fuel calibrations are quite complex, involving MAF, Injector Flow Rate, Volumetric Efficiency, and Closed Loop tables. Practically every aspect of fuel delivery can be manipulated. The ignition calibration is more complicated still, ranging from simple spark advance tables to complicated knock sensor variable attack and decay behaviors depending on severity of detonation. Keep in mind, all fuel and ignition functions are tied to engine and ambient temperatures, and have an interdependent impact on each other.

LS1 Edit is best used on a laptop computer while tuning on a dyno with a wide-band oxygen sensor.

Torque Management

The Gen-III PCM uses a Torque Management system that reduces engine torque just prior to gear changes in cars with automatic transmissions. Considering the questionable durability of this transmission in this application, due care should be exercised when disabling Torque Management.

Companies such as Superchips have partnered with shops that use their Micro-Tuner box to eliminate the need to ship the PCM around the continent for custom tuning. The way it works is this: The shop fills out a form supplying pertinent information about a car's combination. They fax the info to Superchips who then sets to work on the calibration. When complete, the calibration file is placed on the company's ftp site from which the shop can download it to the Micro-Tuner box. From there it is a simple five-minute procedure to load the program into the car's PCM. The obvious advantage is that the car does not become a 3500-pound paperweight while the PCM is en route to and from the programmer. Turn-around times are usually reduced, as well.

LS1 Edit

A powerful tool for the do-it-yourself PCM calibrator is a software and hardware package called LS1 Edit. It allows you to perform nearly any of the modifications a tuner can. It's a comprehensive program with the ability to modify all the relevant maps required to cope with all of the possible significant hardware changes intended for increased performance. While LS1 Edit is clearly useful for such mundane tasks as changing the final drive ratio settings, its true utility is in its advanced calibration capabilities. For example, one excellent feature of LS1 Edit is the ability to graphically display calibrations. This allows the tuner to easily visualize the tune and observe any odd tendencies in the way a car is set up. This function is also very useful in allowing tunes to be overlaid, making direct comparisons easy.

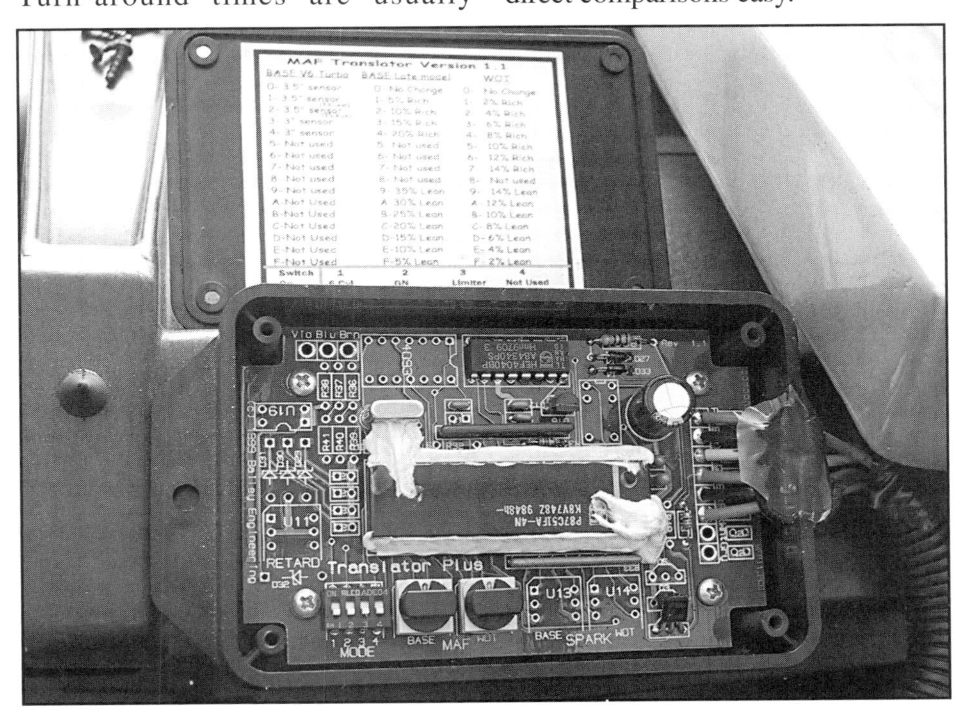

Adjustments made with any tuning device should always be monitored with either a scan tool or on a dyno with a wide-band oxygen sensor.

The Mass Air Flow Translator is a versatile and easy-to-use tool for adjusting the LS1's fuel trim.

Usage—Because the Gen-III PCM uses flash memory, LS1 Edit can't modify PCM code in real time like some aftermarket systems can. After the calibration is complete, it takes less than five minutes to upload the new data tables to the PCM.

At first glance, LS1 Edit may look like a magic wand, bestowing unlimited power with a few clicks of a mouse. Despite the software's sophistication and the relative ease with which PCM parameters can be accessed, it won't make you an expert in PCM calibration. Untold numbers of engines and transmissions have been and will continue to be sacrificed as novice home tuners discover PCM calibration is no cakewalk. LS1 Edit is ideal for the knowledgeable enthusiast looking to gradually modify his car over time. If he knows what he's doing, it will save him the repeated expense of a professional recalibration each time he changes his combination.

MAF Translator

The Translator is a small device that allows adjustment to the fuel delivery of the engine. It manages this by intercepting the signals between the Mass Air Flow meter and the PCM. It can modify these signals to change the fuel delivery rate. The Translator can add or subtract 0-14% fuel from the WOT fuel map of the engine, and 0-20% from the base fuel map. The MAF Translator can also compensate for larger injectors (usually up to 30 lbs-hr.) or otherwise modified combinations without having to purchase custom engine tuning. The MAF translator plugs in between your mass air flow sensor (MAF) and the factory harness. Installation only takes a few minutes. For best and safest results, a wide band oxygen sensor should always be used to tune the PCM in conjunction with this tool.

11 Racing and Transplanting the LS1

RACING

The LS1 is truly a fantastic engine, so it should come as no surprise to find it beneath the hoods of a wide variety of racecars.

C5-R

The first factory-backed race effort was to be in the endurance arena, powering the engine's progenitor, the Corvette. In the autumn of 1997, the C5-R project was launched. The project targeted endurance races ranging from the 12 Hours of Sebring to the 24 Hours of Daytona, all the way to the 24 Heures du Le Mans, with victory at Le Mans being the ultimate goal.

The project got under way using production parts, such as the stock block, cylinder heads, and intake manifold. The rotating assembly was all top shelf, with JE forged pistons, Carillo rods and a billet steel crankshaft. The block was re-sleeved and bored to 4.060-inches bore and the crank stroke was shortened to 3.410-inches. Final displacement was 353 cubic inches and the engine produced 540 horsepower with a production intake manifold and revved comfortably to 7500 rpm. Replacement of the stock plastic manifold with a fabricated aluminum sheet metal piece brought a bump to 580 horsepower. While this was certainly strong, it was deemed insufficient to be competitive.

Thus was born the Step Two aluminum alloy (a.k.a. C5-R) cylinder head. A significant change was a reduction in valve angle to just 11°. This allowed the combustion chamber to be reduced and allowed a significant increase in power. The extra power was nowhere to be found, however, when the C5-R made its racing debut at Daytona in the spring of 1999. That's because the car had to be run with air restrictors there, limiting output to 585 horsepower at 7200rpm. Even with the diminished air requirements, the air filtration system was inadequate and the engines ingested debris. Engine wear quickly became a problem as piston blow-by increased to the point that the crankcase scavenge pumps couldn't keep up. The result was blown crank seals.

Experimentation continued, and a new approach was taken. Contrary to non-restricted engines, where more horsepower is made by increasing

The Corvette C5-R program was conceived to re-establish Chevrolet in world sports car racing at such venues as Daytona, Sebring, and most importantly, the Le Mans 24-Hour race.

The C5-R engine program relies heavily on OEM design. Though utilizing race-specific cylinder heads, block and rotating assembly, the basic architecture matches production engines. Other notable differences are the Kinsler fuel injection system and dry sump oiling system.

engine rpm, the C5-R team set out to make maximum power at the lowest speed possible. In order to do this, stroke was increased to 4.00-inches, making displacement 414 cubic inches, or just under 7 liters. A 6.0-inch connecting rod was used to improve the stroke-to-rod ratio, improving the piston angle around top dead center. The new engine configuration was first raced at Laguna Seca late in the 1999 American Le Mans season. It was making 610 horsepower from 6200 rpm to 6400 rpm, with the rev limiter set at 6800 rpm.

The C5-R block was designed around the production castings, but with several important differences. Among them are thicker aluminum walls surrounding the liners, smaller water jackets, and dry iron liners, which are pressed, rather than cast, in place. Other trick pieces include a dry sump oil pan (which can be used on the stock engine) and billet steel main caps. The timing chain has higher quality sprockets and a beefier chain, and the pushrods were upgraded to .375-inch diameter chro-moly pieces. The heads use titanium valves and copper-beryllium valve seats. Fuel is supplied to the Kinsler velocity stack fuel injection system by an in-tank electric pump at a pressure of 72 psi (5-bar).

GM Powertrain's John Juriga is very upbeat on the C5-R project, stating: "Things have gone really well, and we've learned some things that we can apply to the production engines. There are things we are working on now that will be directly attributable to C5-R program."

ASA Racing

Stock car racing is a completely different world from that of endurance road racing. When the American Speed Association announced its intention to mandate a spec-engine, GM was quick to see to it that the LS1 was their first choice. The relationship started in 1998, when General Motors and ASA embarked on a joint project to introduce electronic engine management to oval track racing. The

Though plagued by minor new-car gremlins during its first season, the C5-R program has gone on to dominate nearly every race it has attended —including Le Mans.

plan was to utilize a high-performance, factory-built Gen-III powerplant.

ASA initially ordered 300 engines from GM in 1999. The engines were assembled at GM's Romulus, Michigan, engine plant over the span of only three hours. Once built, the engines were shipped to Lingenfelter Performance Engineering (LPE) in Decatur, Indiana, where each engine was modified, tested and sealed before being sent to the teams which would then install them in the racecars.

The LS1s used in ASA are built on the production engine assembly line in Romulus, Michigan.

Modifications performed at LPE include a more aggressive camshaft (525-inches lift, 226°/236° duration, 110° LSA) and more durable valve springs. Each engine uses a dry-sump oil system, and new rod bolts. The PCM is a special calibration with different values for fuel and spark. The engine was rated at 430 horsepower at 6200 rpm and 430 lb.-ft. of torque at 4800 rpm. ASA mandated the use of this engine for the 2000 AC-Delco Series.

For GM, this was a great chance to learn about the engine's performance and durability under extreme racing conditions, as everything they learned on the track could be applied to production engines. Over the course of the first two seasons, ASA drivers logged 270,600 miles on the engines. Those two seasons showed very few failures.

This ASA-GM partnership has also had a tremendous economic impact on the teams competing in ASA. The GM engines are sold to the teams for $12,000 each. The engines have been reliable enough that several teams have run a full 20-event season utilizing a single engine. Prior to the LS1, the teams relied on aftermarket V6 power plants that often cost $30,000 or more. It has been

An LS1-powered ASA Monte Carlo Adoption by the American Speed Association as its spec engine has proven valuable to GM in evaluating the engine's long-term durability in extremely strenuous conditions.

Lingenfelter Performance Engineering modifies each engine with a higher lift camshaft and a dry-sump oiling system.

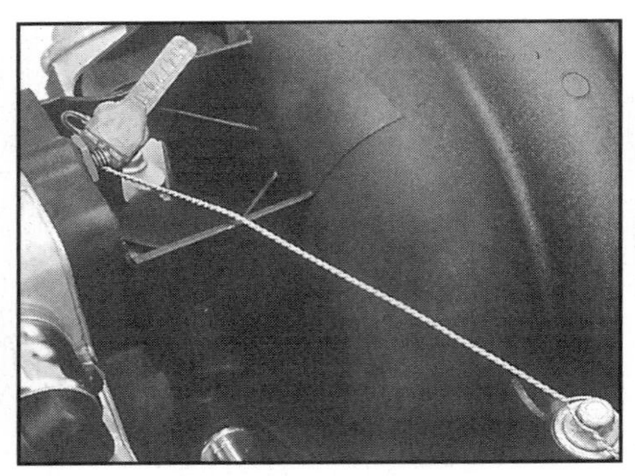
After modification, the engines are sealed and shipped to the race teams.

The LS1s in ASA use 85mm mass airflow sensors with integrated intake air temperature sensors.

suggested that this program has reduced the teams' engine budgets by as much as $100,000 per season.

John Juriga says he is quite satisfied with GM's involvement in the ASA program. "We're very happy with the way it has gone. The engine has really impressed us with its durability. It gave us a chance to run the engines at a higher power level than we may have tested for. The lower end has been exceptionally durable. Given proper coolant and fresh oil, the engines seem happy to run and run. Racing allowed us to find a potential weak spot in the LS1. Due to failures in ASA, we changed the way we manufactured the rod bolts for the production engines. We now heat treat the bolts and then roll threads to work harden the material. That's an excellent example of a lesson learned racing that went into production. Overall, I think the durability of the LS1 in ASA racing is a great testament to durability of the design."

LS1s are becoming a common sight between the fenders of classic musclecars like this '68 Camaro.

TRANSPLANTS

The transplanting of electronically controlled fuel injected engines into classic musclecars has been going on for nearly twenty years now. But it is in the last five to ten years that this movement has really picked up momentum. Thanks in part to the killer engines coming out of Detroit, and to innovative companies like Painless Performance Products, which have created an entire catalog full of products that make the process practically plug-n-play.

All of the attributes that make the LS1 such a great engine in the F-bodies and Corvettes also make it highly desirable to those looking for an alternative engine for their classic muscle car or street rod. Not only do you get marvelous street manners and fuel efficiency, but you get the prodigious power the all-aluminum stormer is known for. While this is cool in and of itself, when is the last time you saw one dropped into a first-Gen with the sole intention of dragstrip supremacy? Leave it to those crazy Canadians at Agostino Racing Engines to be the first to light the fuse on an all-out quarter-mile assault. Their install begins on the next page.

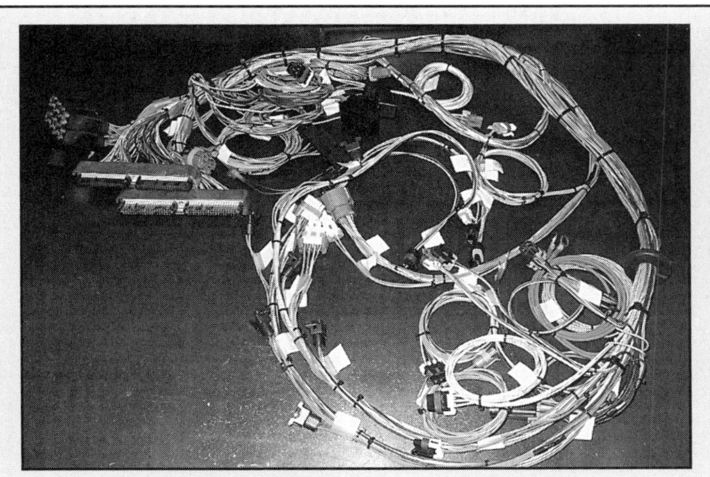

One of the most important aspects of the install was the use of Painless Wiring's LS1 engine harness. It allows the transplant of an LS1 engine into nearly anything with four wheels and a place to sit. Like all Painless harnesses, the LS1 harness uses high temp, lightweight TXL wire for optimum performance. A Vehicle Anti-Theft System (VATS) module is incorporated into the harness to bypass the vehicle anti-theft system, so no reprogramming of the computer is required.

This install used an F-body alternator bracket (which mounts the alternator low on the driver's side) to keep the install as tidy as possible. To do this, the cross-member had to be notched. Use of a top-mounted Corvette or truck style alternator bracket would eliminate this necessity.

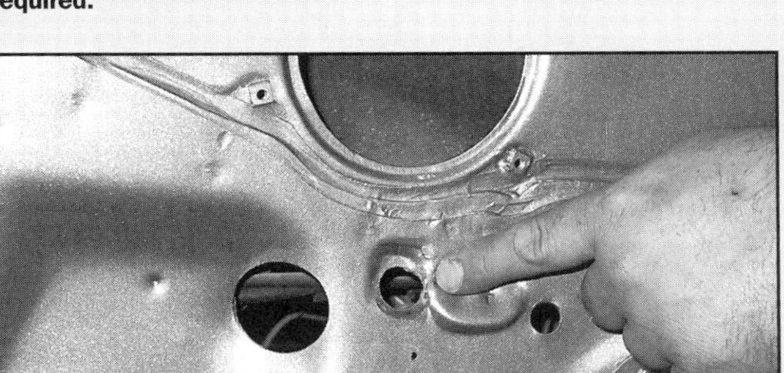

The firewall required only minor modification in the way of a pair of drilled holes. The small hole on the right is the throttle cable hole that has been bored to 9/16 inches in order to accommodate the late-model F-body cable. The 1-5/8-inch hole to the left was bored for the Painless harness to pass through. Painless supplies a matching grommet with the harness to seal things up.

The stock steering works fine, but is in the way of the LS1 F-body oil pan sump. Therefore, a stock F-body oil pan was modified to clear the rear-steer Camaro componentry by shortening the sump.

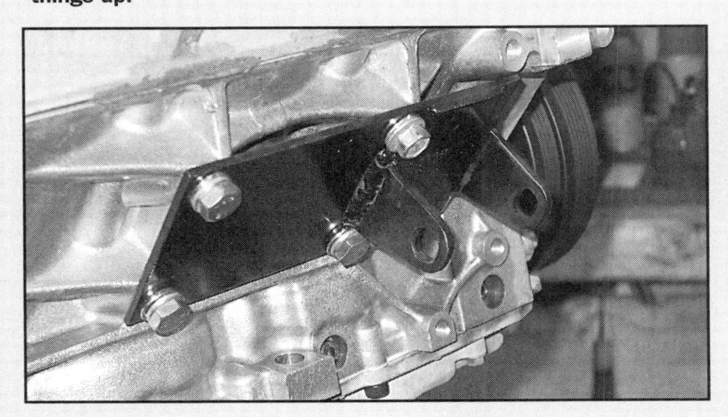

Because this car was built for maximum effort quarter-mile assaults, solid motor mounts were used. Their low profile allows the LS1 to sit quite low in the chassis. Agostino Racing Engines offers these conversion motor mounts in both solid and urethane versions.

Mating the LS1 to a Turbo 400 transmission is no problem (it bolts right up), but the converter will need a new pilot bushing installed.

As you can see, the LS1 sits very low and close to the firewall. This dramatically improves the car's center of gravity.

With the engine in place, the notched oil pan provides plenty of clearance for the steering gear.

A 135-amp alternator from a Gen-III truck engine was chosen for its extra charging capabilities. There is plenty of air space in the pocket surrounding it to allow for proper cooling.

No one made headers for this engine conversion at the time it was done, so custom pieces were built. These step headers transition from 1 3/4-inch to 1 7/8-inch diameter and have 3-1/2-inch collectors with oxygen sensor provisions.

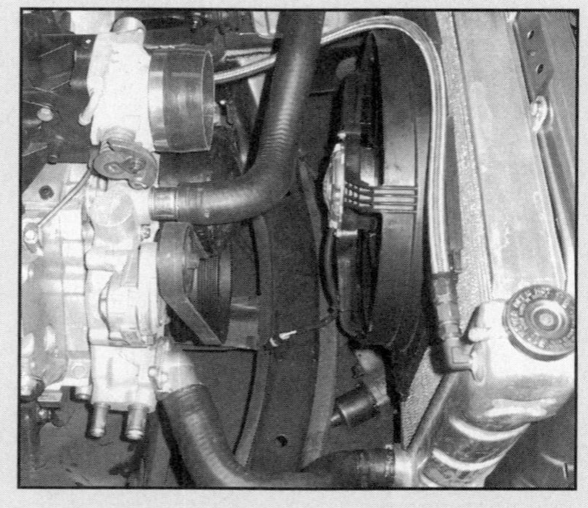

The aluminum radiator was a leftover from the car's former big-block. It was modified with the simple addition of a –6 AN fitting to mate to the LS6 cooling rail that is required by the LS6 intake manifold. A Corvette upper and a slightly trimmed big-block Camaro lower radiator hose connects everything else.

A mandrel-bent piece of 4-inch tubing mates the throttle body to this simple and effective air induction system. The 14-inch K&N filter is mated to an LS6 mass airflow meter.

Once most of the hardware was bolted in place, it was time to feed the harness through the firewall and to the engine room. Each weatherpack connector is clearly labeled and of the highest quality. Painless supplies plenty of wire length to accommodate creative routing schemes.

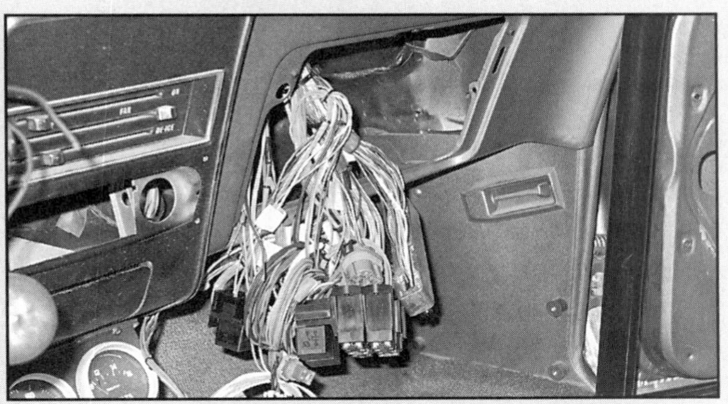

Because this car will not be subject to emissions inspections, a few items such as the EGR system and AIR pump were omitted. Rather than cut the harness, these connectors were simply tied back to the harness beneath the dashboard. Notice the relays and fuse block are also kept within the safe confines of the car's interior. The transmission control harness (with provisions for both a 4L60-E and T-56) are tucked away for safe keeping as well.

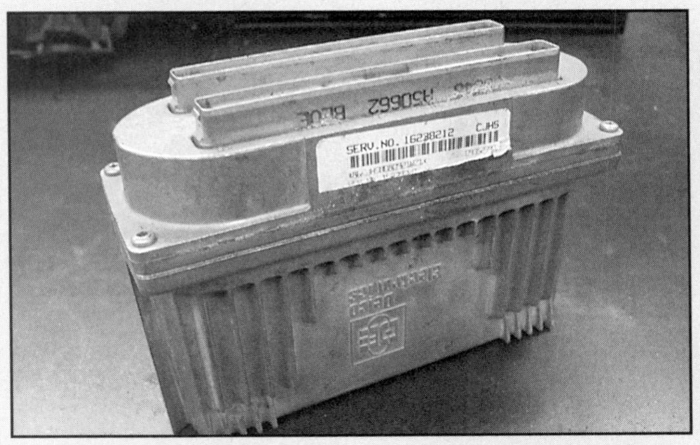

The Painless Wiring LS1 harness requires use of the PCM from a '98 or '99 Camaro (GM P/N 16238212). These are available from any GM dealer or even at some salvage yards. If you are installing a stock engine, reprogramming the PCM is unnecessary.

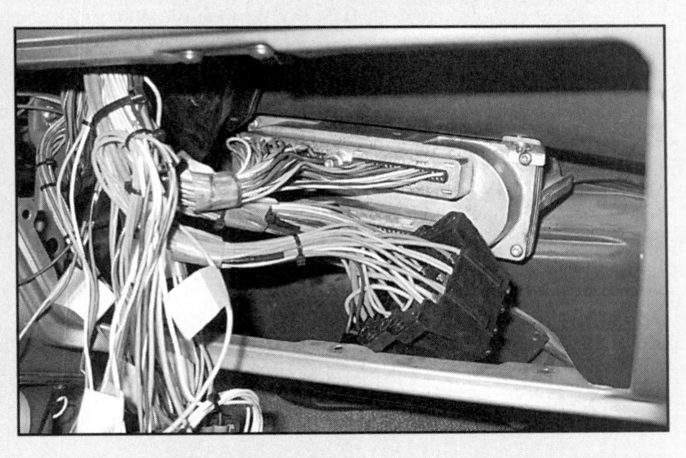

The creation of a couple of simple brackets allowed the PCM to be hung safely behind the glove box.

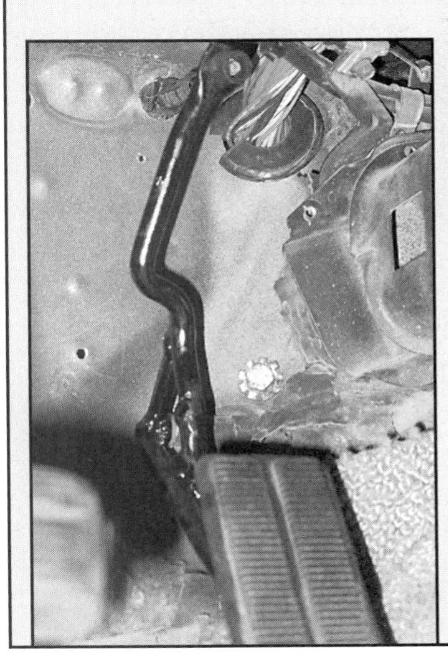

The throttle pedal was modified by adding the lever from a late-model car to the original '67 actuator. With the pedal mounted, you can see it interfaces perfectly with the late-model throttle cable.

The finished install looks innocent enough, but just wait until the two-stage nitrous system is installed.

Aeromotive, Inc.
5400 Merriam Drive
Merriam, KS 66203
913/647.7300
www.aeromotiveinc.com

Agostino Racing Engines (ARE)
1652A Bayly Street
Pickering, Ontario, Canada L1W 1L9
905/420.9195
www.agostino-racing.com

Automotive Racing Products (ARP)
1863 Eastman Avenue
Ventura, CA 93003
805/339.2200
www.arp-bolts.com

Accessible Technologies, Inc. (Procharger)
14801 West 114th Terrace
Lenexa, KS 66215
913/338.2886
www.procharger.com

B&B Performance Exhaust
23045 North 15th Ave.
Phoenix, AZ 85027
623/581.7600
www.bbfabrication.com

Borla Performance Industries
5901 Edison Drive
Oxnard, CA 93033
877/462.6752
www.borla.com

Carputing (LS1 Edit)
http://carputing.tripod.com

Childs & Albert Inc.
24849 Anza Drive
Valencia, CA 91355
661/295.1900
www.childs-albert.com

Cometic Gasket
8090 Auburn Road
Concord, OH44077
440/354.0777
www.cometic.com

Comp Cams
3406 Democrat Road
Memphis, TN 38118
907/795.2400
www.compcams.com

Corsa Performance
140 Blaze Industrial Parkway
Berea, OH 44017
440/891.0999
www.corsaperf.com

Crane Cams
530 Fentress Boulevard
Daytona Beach, FL 32114
386/252-1151
www.cranecams.com

Doug Rippie Motorsports (DRM)
5767 State Highway 55 SE
Buffalo, MN 55313
763/477.9272
www.dougrippie.com

Fastchip
1114 West 41st Street
Tulsa, OK 74107
918/446.3019
www.fastchip.com

Finish Line Performance
10S 059 Schoger Road
Naperville, IL 60564
630/820.9294
www.flp2win.com

GM Performance Parts
800/GM.USE.US
www.spoperformanceparts.com

Halltech
9402 Gulstrand Circle
Huntington Beach, CA 92646
714/963.3000
www.corvettec5.com

Holley Performance Products PO Box 10360
Bowling Green, KY 42102
www.holley.com

Hypertech, Inc.
3215 Appling Road
Bartlett, TN 38133
901/382.8888
www.hypertech.com

Jesel Valvetrain Innovation
1985 Cedar Bridge Avenue
Lakewood, NJ 08701
712/901.1800
www.jesel.com

K&N Engineering
PO Box 1329
1455 Citrus Avenue
Riverside, CA 92502
909/826.4000
www.knfilters.com

Katech Engine Development
24324 Sorrentino Court
Clinton Township, MI 48035
810/791.4120
www.katechengines.com

King Engine Bearings, Inc.
371 Little Falls Road
Cedar Grove, NJ 07009
973/857.0705
www.kingbearings.com

Lingenfelter Performance Engineering (LPE)
1557 Winchester Road
Decatur, IA 46733
260/724.2552
www.lingenfelter.com

LS1 Motorsports
397 Highway 62
Charlestown, IA 47111
812/256.6400
www.ls1motorsports.com

Magna Charger
3172 Bunsen Avenue, Unit D,
Ventura, CA 93003
805/289-0044
www.magnacharger.com

Manley Performance Products
1960 Swarthmore Avenue
Lakewood, NJ 08701
732/905.3366
www.manleyperformance.com

Motorsport Technologies, Inc.
12500 Oxford Park Drive
Houston, TX 77082
281/870.8787
www.motorsporttech.com

MSD Ignition
1490 Henry Brennan Drive
El Paso, TX 79936
915/857.5200
www.msdignition.com

Nitrous Warehouse
192 Oak Ridge Street
Cleveland, TX 77327
281/592.7664
www.nitrouswarehouse.com

Oliver Racing Parts
1025 Clancy Ave. NE
Grand Rapids, MI 49503
616/451.8333
www.oliver-rods.com

Jim Pace GM Parts Warehouse (Pace Performance Parts)
430 Youngstown-Warren Road
Niles, OH 44446
800/748.3791
www.paceparts.com

Painless Performance Products
9505 Santa Paula Drive
Fort Worth, TX 76116
888/350.6588
www.painlesswiring.com

Racing Engine Valves (REV)
4704 NE 11th Ave.
Fort Lauderdale, FL 33334
954/772.6060
www.revvalves.com

Random Technology
4430 Tuck Road
Loganville, GA 30052
770/554.4242
www.randomtechnology.com

Ross Racing Pistons
625 South Douglas
El Segundo, CA90245
310/536.0100
www.rosspistons.com

SCAT Enterprises, Inc.
1400 Kingsdale Avenue
Redondo Beach, CA 90278
310/370.5501
www.scatcrankshafts.com

SLP Performance Parts
1501 Industrial Way N
Toms River, NJ 08755
732/349.2109
www.slponline.com

Suncoast Creations
3014 59th Avenue East
Bradenton, FL 34203
888/754.8900
http://www.suncoastcreations.com

Superchips, Inc.
134 Baywood Avenue
Longwood, FL 32750
800/898.2447
http:/www.superchips.com

Texas Nitrous Technology
194 FM 346 E.
Tyler, TX 75703
903/839.5498
www.nitrous-power.com

Thunder Racing
6960 North Merchant Court
Baton Rouge, LA 70809
225/754.7223
www.thunderracing.com

TTS Power Systems
1280 Kona Drive
Compton, CA 90220
310/669.8101
www.ttspowersystems.com

Vortech Superchargers
1650 Pacific Avenue
Channel Islands, CA 93033
805/247.0226
www.vortechsuperchargers.com

APPENDICES

Combination 1: Bolt-Ons + TNT Nitrous Oxide

Car: 1999 Z28
Transmission: 6-speed
Clutch: McLeod
Gears: 4.10
Induction: MTI lid, K&N filter, ported stock MAF and throttle body
Manifold: LS6
Heads: stock LS1
Cam: MTI T1: 221°/221° duration, .544/.544-inches lift, 112° LSA
Headers: Grotyohann 1.75-inch with 3-inch collectors and off-road Y-pipe
Exhaust: Borla, no plate
Nitrous System: TNT Power Ring
Fuel Injectors: stock
Fuel Pump: stock in-tank with NOS in-line

Naturally Aspirated:
Horsepower: 376 rwhp
Torque: 374 ft.-lbs.

75 HP Nitrous Injection
Horsepower: 478 rwhp
Torque: 558 ft.-lbs.

150 HP Nitrous Injection
Horsepower: 524 rwhp
Torque: 629 ft.-lbs.

DYNOJET Performance Evaluation Program

DYNORUN.005 150 SHOT, RICHRO 11/7/01 11:27:12 AM
DYNORUN.002 75 SHOT RO 11/7/01 10:36:48 AM
DYNORUN.001 RO 11/7/01 10:30:16 AM

Elite Autosport 612-521-0550

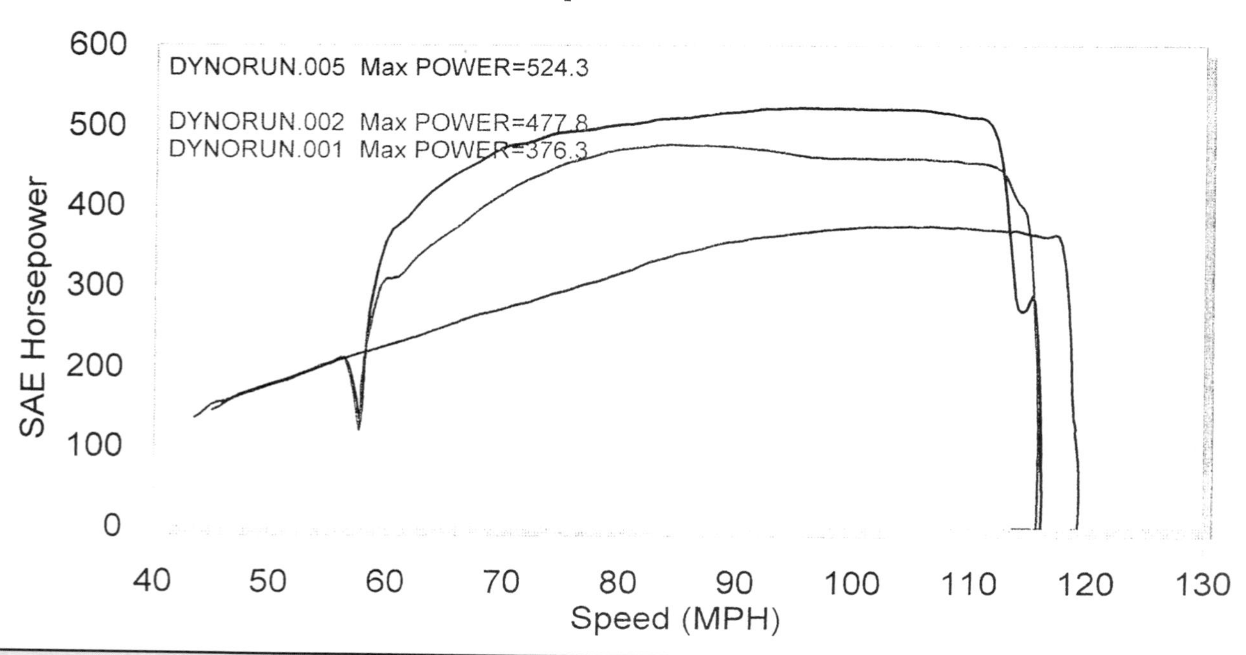

DYNOJET Performance Evaluation Program

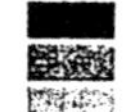

DYNORUN.005 150 SHOT, RICHRO 11/7/01 11:27:12 AM
DYNORUN.002 75 SHOT RO 11/7/01 10:36:48 AM
DYNORUN.001 RO 11/7/01 10:30:16 AM

Elite Autosport 612-521-0550

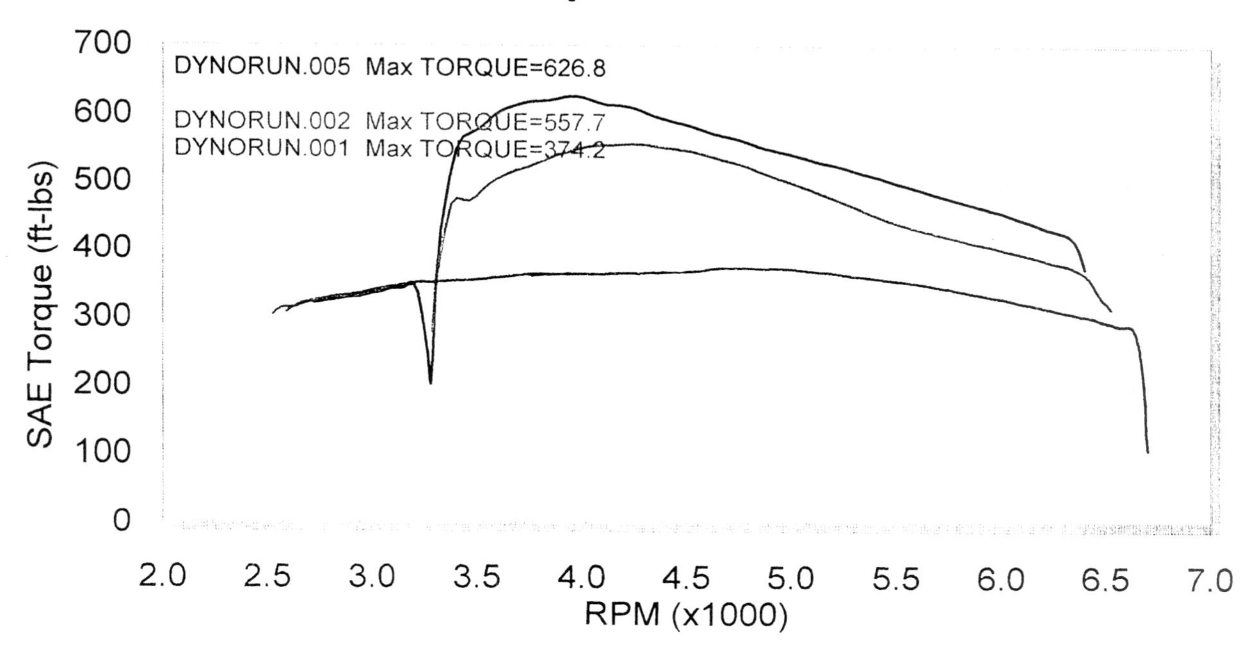

Combination 2: ARE Stage 2 Heads & Cam

Car: 1998 Z28
Transmission: 6-speed
Clutch: McLeod
Gears: 4.11 (12-bolt rear)
Induction: MTI lid, K&N filter, ported stock MAF and throttle body
Manifold: LS6
Heads: ported LS1, 2.055/1.60 valves,
Cam: Lunati 222/230° duration, .534/.544-inches lift, 114° LSA
Headers: Grotyohann 1.75-inch with 3-inch collectors and off-road Y-pipe
Exhaust: B&B Tri-Flow
Injectors: stock
Fuel pump: stock

Horsepower: 411 rwhp
Torque: 388 ft.-lbs.

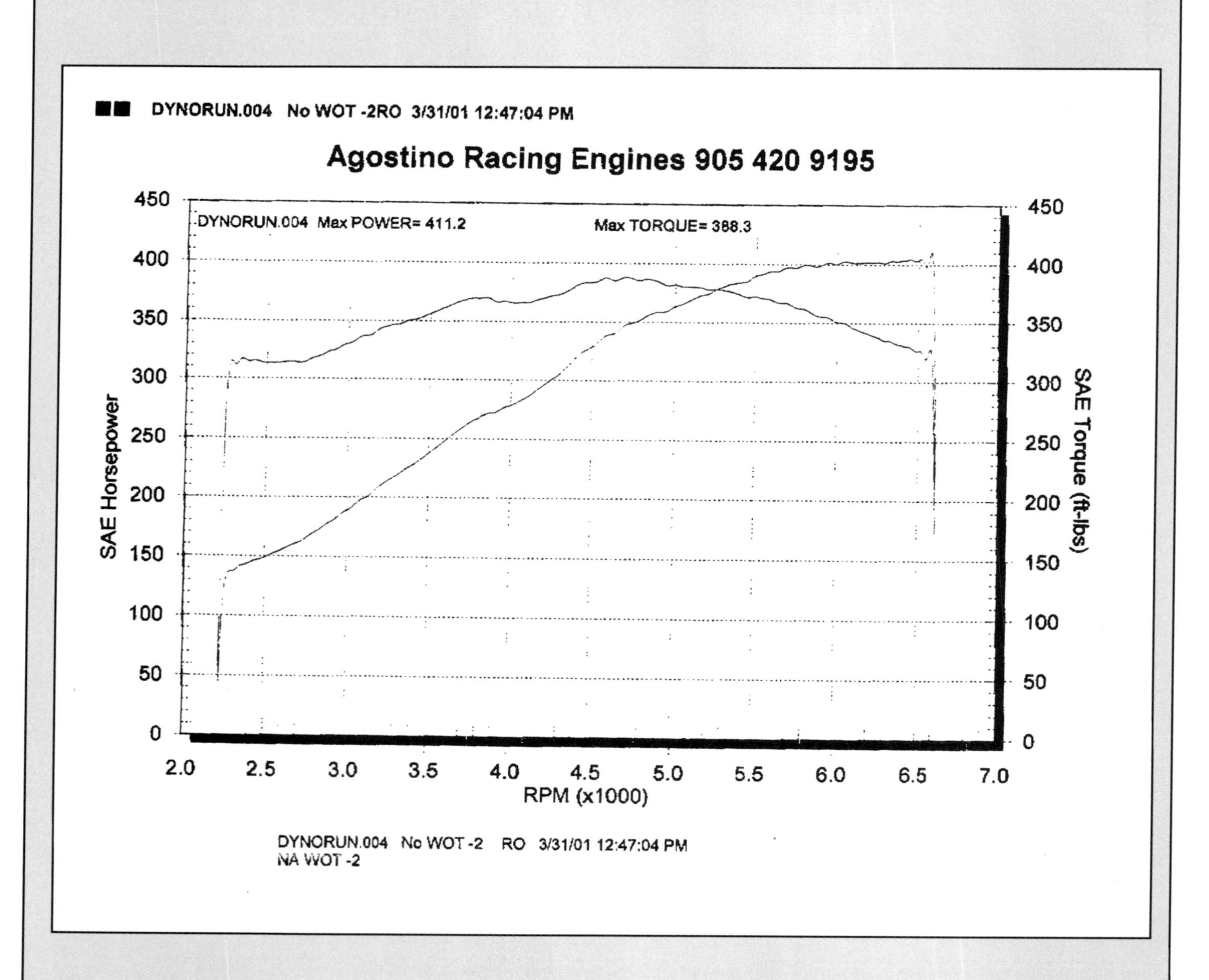

Combination 3: Bolt-Ons + LS1 Motorsports Turbo

Car: 2001 Trans Am
Transmission: 6-speed
Clutch: McLeod
Gears: 3.42
Induction: LS1 Motorsports T-63 turbo, ported stock MAF and throttle body
Manifold: LS6
Heads: stock LS1
Cam: stock
Headers: LS1 Motorsports turbo
Exhaust: 3-inch downpipe with Borla, no plate
Boost: 9 psi

Fuel Injectors: 37 lbs.-hr
Fuel Pump: Walbro 340 in-tank

7 psi:
Horsepower: 534 rwhp
Torque: 678 ft.-lbs.

9 psi:
Horsepower: 534 rwhp
Torque: 678 ft.-lbs.

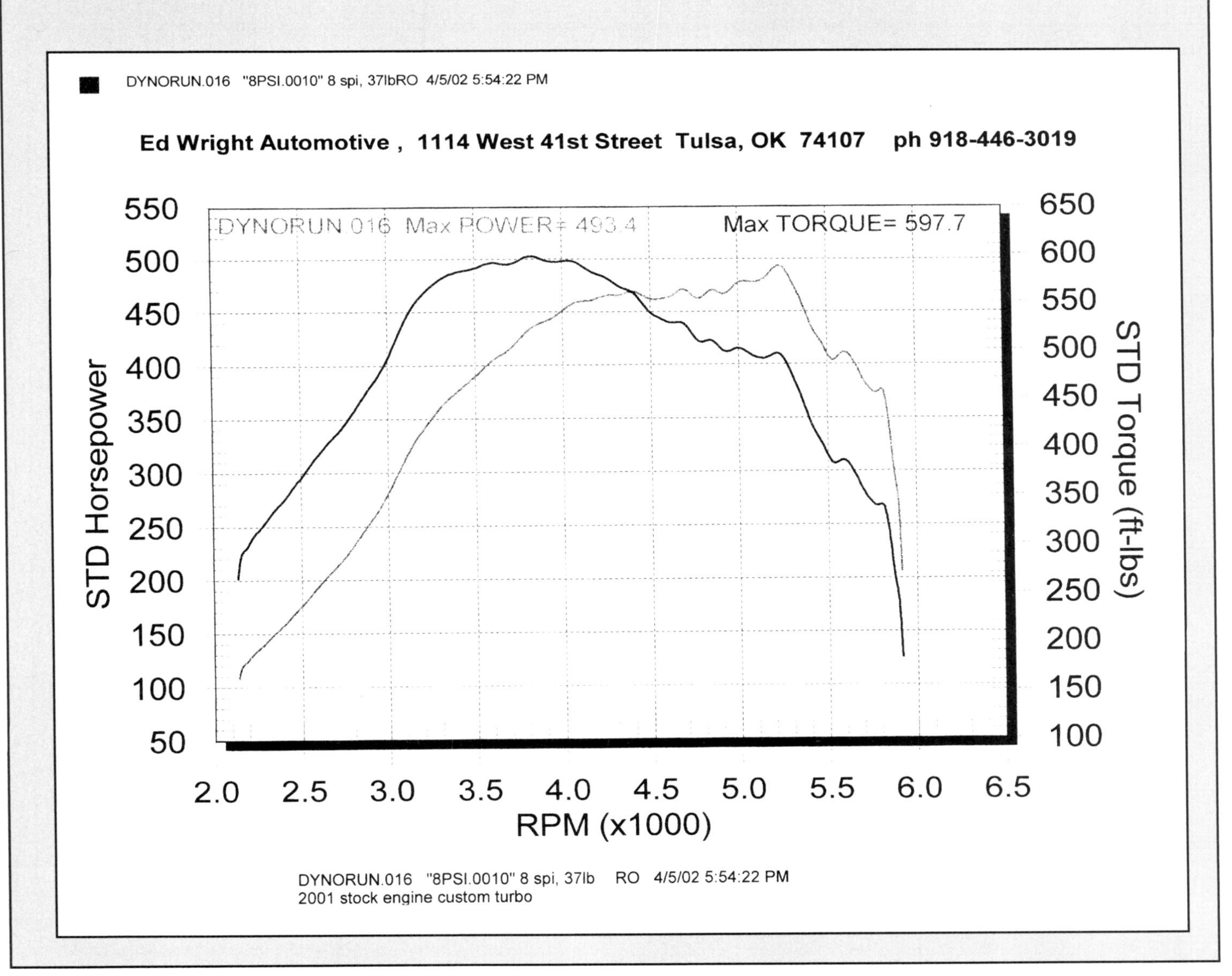

Combination 4: MTI 382 ci Stroker

Car: 1998 Z28
Displacement: 382 ci
Compression: 11.25:1
Transmission: 6-speed
Converter: McLeod
Gears: 3.73
Induction: MTI lid, K&N filter, stock MAF and throttle body
Manifold: LS1
Heads: ported LS1, 2.08/1.60 valves, springs, titanium retainers
Cam: MTI hydraulic roller 221°/221° duration, .558/.558-inches lift, 116° LSA
Shortblock: Wiseco pistons, Lunati rods and crankshaft
Headers: Grotyohann 1.75-inch with Y-pipe
Exhaust: Borla, middle plate
Injectors: stock
Fuel Pump: stock

Horsepower: 440 rwhp
Torque: 436 ft.-lbs.

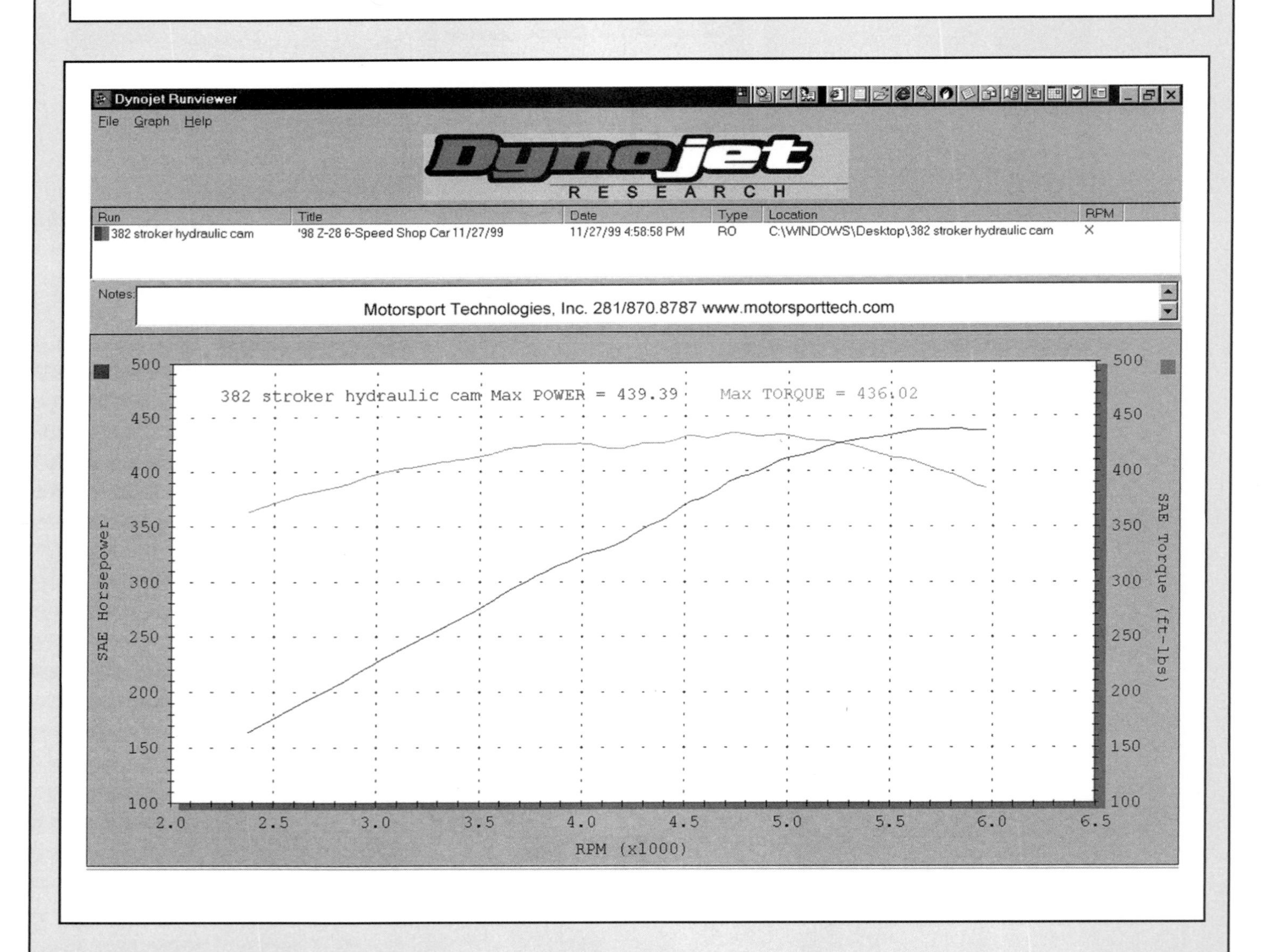

Combination 5: ARE 393 ci Big-Bore with Solid Roller

Car: 1998 Z28
Displacement: 392 ci
Compression: 12:1
Transmission: 200-4R
Converter: Yank 4600 rpm stall speed
Gears: 4.11
Induction: MTI lid, K&N filter, ported stock MAF and throttle body
Manifold: LS6
Heads: ported LS6, Manley 2.08/1.60 valves, springs, titanium retainers
Cam: Crane solid roller 252°/260° duration, .680/.707-inches lift, 110° LSA
Shortblock: Ross pistons, Lunati rods, offset ground crank
Headers: Pace 1.75- to 1.88-inch stepped, 3.5-inch collectors
Exhaust: 3.5-inch, 2-chamber Flowmaster mufflers
Injectors: Lucas 43.5 lbs.-hr. (@ 3-bar)
Fuel Pump: Walbro 340 in-tank

Horsepower: 469 rwhp
Torque: 461 ft.-lbs.

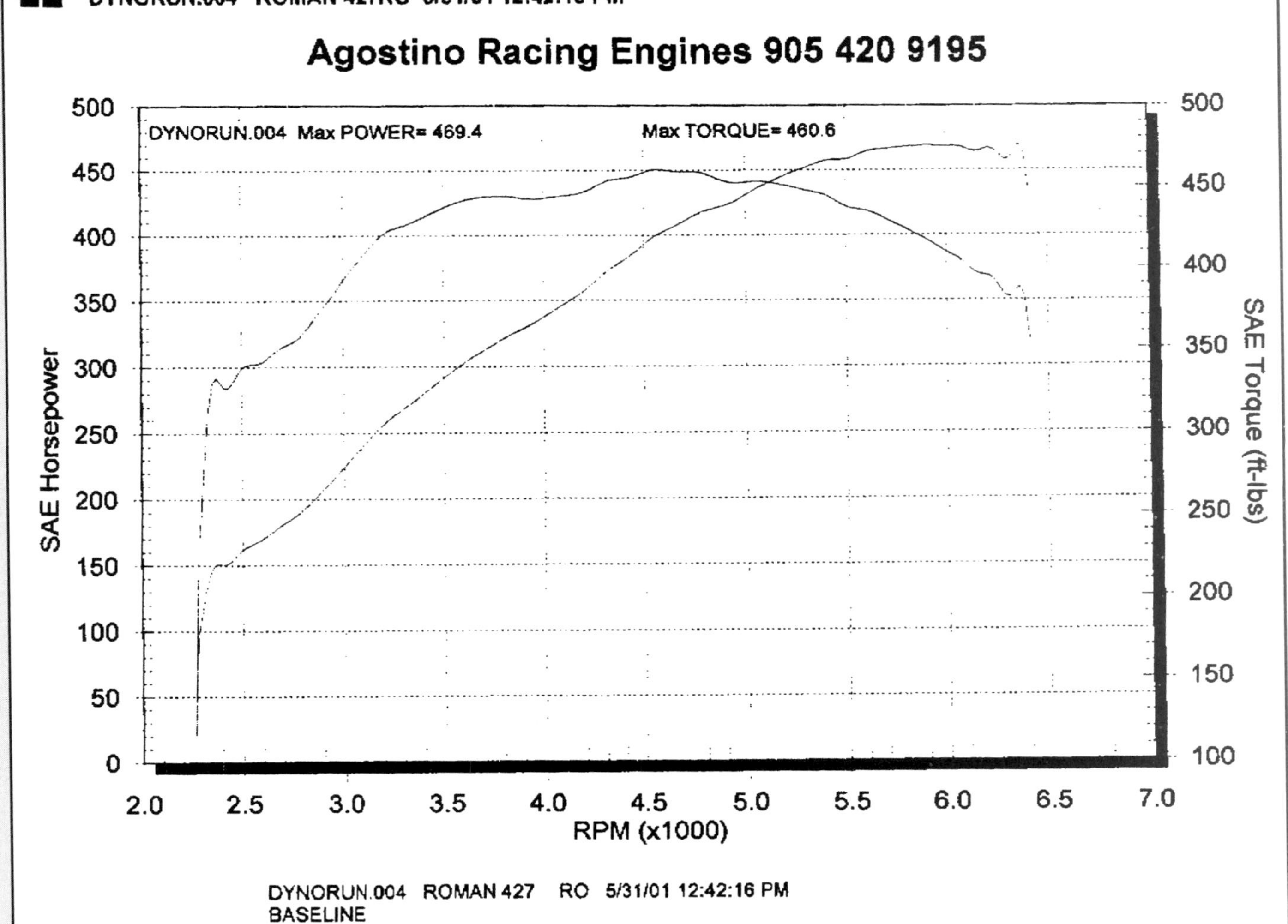

Combination 6: MTI 422 ci Super Stroker with Solid Roller

Car: 1998 Corvette
Displacement: 422 ci
Compression: 13:1
Transmission: 6-speed
Clutch: McLeod twin disc
Gears: 3.42
Induction: MTI ram air, K&N filter, LS6 MAF and ported throttle body
Manifold: custom sheet metal
Heads: ported LS6, 2.08/1.60 valves, titanium retainers

Cam: Crane solid roller 250°/260° duration, .650/.650-inches lift, 112° LSA
Shortblock: Wiseco pistons, Lunati billet rods and crankshaft
Headers: 1.75-inch stepped, 3-inch collectors and X-pipe
Injectors: 42 lbs.-hr.
Fuel Pump: stock

Horsepower: 533 rwhp
Torque: 495 ft.-lbs.

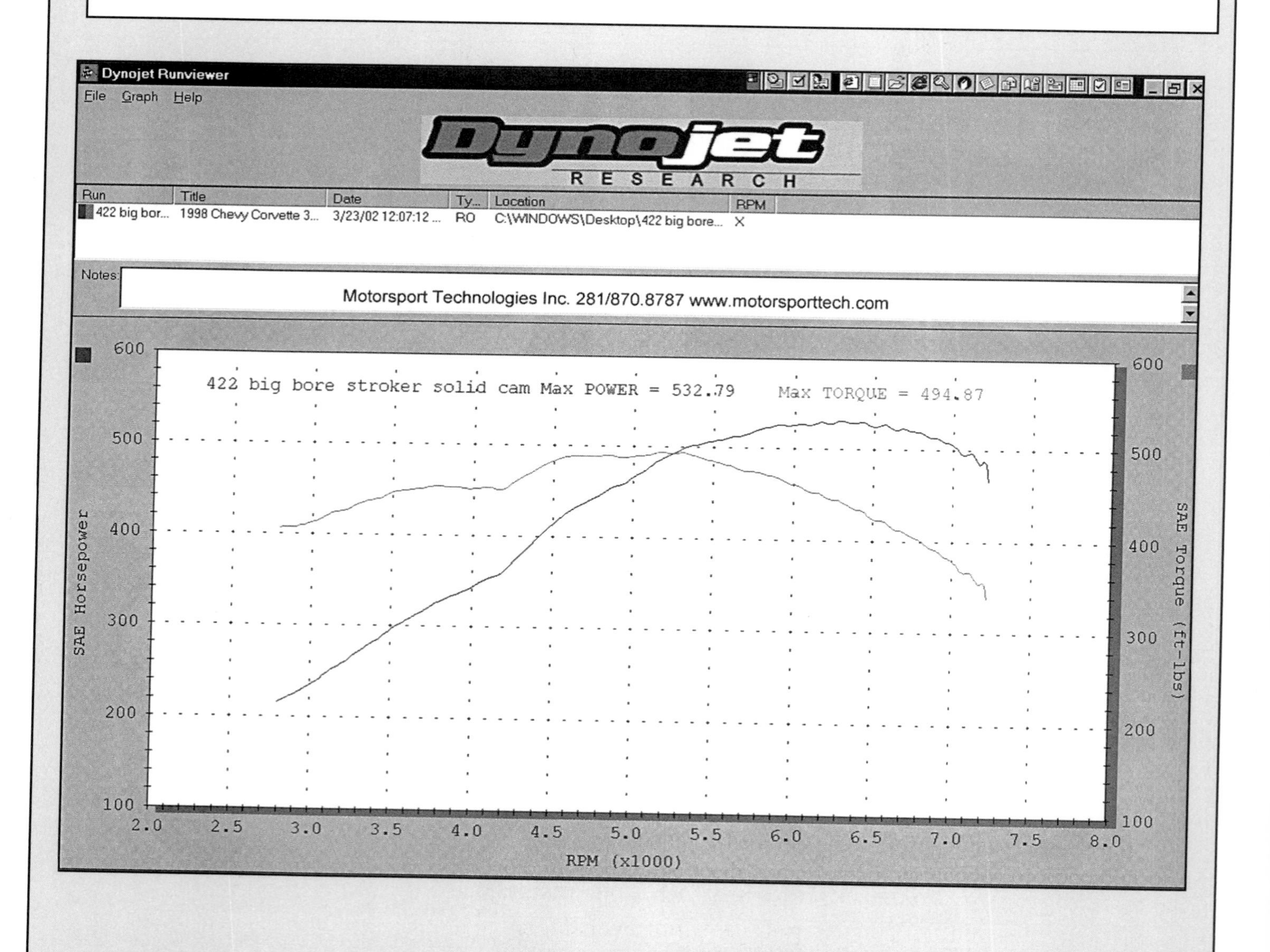

Combination 7: LS1 Motorsports 404 ci Stroker + Turbo

Car: 1999 Z28
Displacement: 404 ci
Compression: 8.5:1
Transmission: Turbo 400
Converter: Pro Yank 3500 rpm stall
Gears: 3.50 (9-inch rear)
Induction: T-88 Turbo, Spearco front-mount intercooler, ported throttle body, SLP 85mm MAF
Manifold: LS6
Heads: ported LS1, 2.08/1.60-inch valves
Cam: Custom hydraulic with 589/612-inches lift
Shortblock: Lunati Pro Billet crank and rods, Ross forged pistons, custom main studs, ARP Head studs, O-ringed block

Exhaust: 3-inch downpipe, 4-inch Mufflex exhaust with Borla XR1 4-inch muffler
Injectors: 95 lbs.-hr.
Fuel Pump: Aeromotive external pump, -10 feed line, -6 return line, Aeromotive 1:1 boost sensitive regulator

10 psi:
Horsepower 608 rwhp
Torque: 640 ft.-lbs.

15 psi:
Horsepower 709 rwhp
Torque: 700 ft.-lbs.

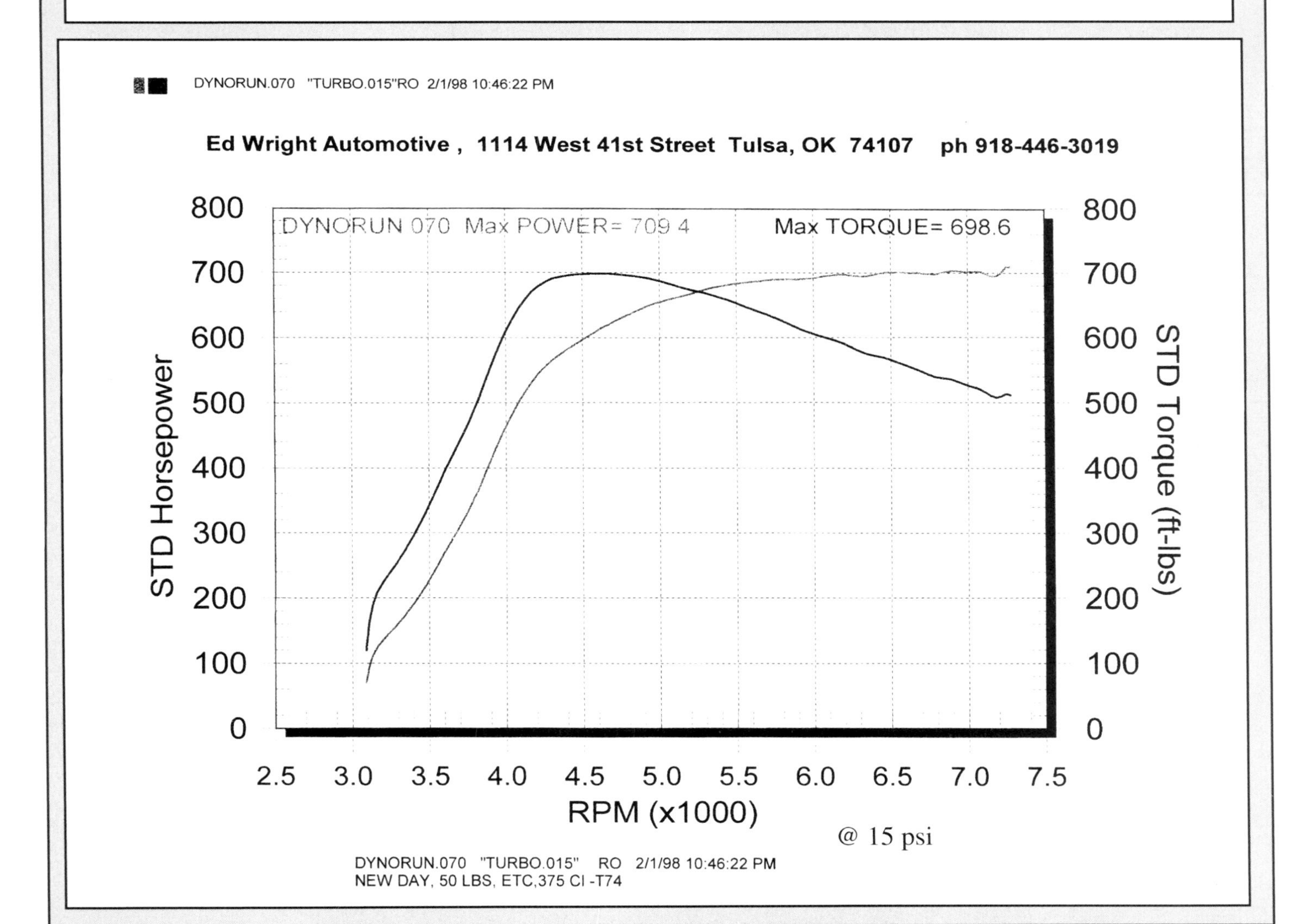

NOTES

Engine Specifications (LS1)

General Data

Engine Type:	V8
Displacement:	5.7L/5665cc/346 CID
Bore:	99.0 mm/3.898 in.
Stroke:	92.0 mm/3.622 in.
Compression Ratio:	10.1:1
Firing Order:	1–8–7-2-6-5-4-3
Spark Plug Type:	41–931
Spark Plug Gap:	1.524 mm/0.06 in.

Lubrication System:	5.678 Liters/6.0 Quarts
Oil Capacity (without Oil Filter Change)	6.0 Quarts/5.7 Liters
Oil Capacity (with Oil Filter Change)	6.5 Quarts/6.151 Liters
Oil Pressure (Minimum—Hot)	
1000 engine rpm:	41.4 kPa/6.0 psig
2000 engine rpm:	124.11 kPa/18.0 psig
4000 engine rpm:	165.48 kPa/24.0 psig
Oil Type: Mobil(E) 5W–30 Synthetic or Equivalent	

Dimensions and Clearances

Camshaft

Camshaft End Play:	0.025–0.305 mm/0.001–0.012 in.
Camshaft Journal Diameter:	54.99–55.04 mm/2.164–2.166 in.
Camshaft Journal Diameter Out-of-Round:	0.025 mm/0.001 in.
Camshaft Lobe Lift (Intake):	7.04 mm/0.277 in.
Camshaft Lobe Lift (Exhaust):	7.13 mm/0.281 in.
Camshaft Runout (Measured at the Intermediate Journals):	0.05 mm/0.002 in.

Connecting Rod

Connecting Rod Bearing Bore Diameter:	56.505–56.525 mm/2.224–2.225 in.
Connecting Rod Bearing Bore Out-of-Round (Production):	0.004 mm/0.00015 in.
Connecting Rod Bearing Bore Out-of-Round (Service Limit)	0.008 mm/0.0003 in.
Connecting Rod Bearing Clearance (Production):	0.015–0.063 mm/0.0006–0.00248 in.
Connecting Rod Bearing Clearance (Service Limit):	0.015–0.076 mm/0.0006–0.003 in.
Connecting Rod Side Clearance:	0.11–0.51 mm/0.00433–0.02 in.

Crankshaft

Crankshaft Bearing Clearance (Production):	0.018–0.054 mm/0.0007–0.00212in.

Engine Specifications, continued

Crankshaft Connecting Rod Journal Dia. (Production):	53.318–53.338 mm/2.0991–2.0999 in.
Crankshaft Connecting Rod Journal Diameter (Service Limit):	53.308 mm (Minimum)/2.0987 in. (Minimum)
Crankshaft Connecting Rod Journal Taper (Production):	0.005 mm/0.0002 in. (Max. for 1/2 of Journal Length)
Crankshaft Connecting Rod Journal Taper (Service Limit):	0.01 mm (Maximum)/0.0004 in. (Maximum)
Crankshaft Connecting Rod Journal Out-of-Round (Production):	0.005 mm/0.0002 in.
Crankshaft Connecting Rod Journal Out-of-Round (Service Limit):	0.01 mm/0.0004 in.
Crankshaft End Play:	0.04–0.2 mm/0.0015–0.0078 in.
Crankshaft Main Journal Diameter (Production):	64.993–65.007 mm/2.558–2.559 in.
Crankshaft Main Journal Diameter (Service Limit:)	64.993 mm (Minimum)/2.558 in.(Minimum)
Crankshaft Main Journal Out-of-Round (Production)	0.003 mm/0.000118 in.
Crankshaft Main Journal Out-of-Round (Service Limit):	0.008 mm/0.0003 in.
Crankshaft Main Journal Taper (Production):	0.01 mm/0.0004 in.
Crankshaft Main Journal Taper (Service Limit):	0.02 mm/0.00078 in.
Crankshaft Reluctor Ring Runout (Measured 1.0 mm/0.04 in.Below Tooth Diameter):	0.25 mm/0.01 in.
Crankshaft Runout (at Rear Flange):	0.05 mm/0.002 in.
Crankshaft Thrust Wall Runout:	0.025 mm/0.001 in.
Crankshaft Thrust Wall Width (Production):	26.14–26.22 mm/1.029–1.032 in.
Crankshaft Thrust Wall Width (Service):	26.32 mm (Maximum)/1.036 in. (Maximum)

Cylinder Bore

Cylinder Bore Diameter:	99.0–99.018 mm/3.897–3.898 in.
Cylinder Bore Taper Thrust Side:	0.018 mm (Maximum)/0.0007 in. (Maximum)

Cylinder Head

Cylinder Head Engine Block Deck Flatness (Measured within a 152.4 mm/6.0 in. area):	0.11 mm/0.004 in.
Cylinder Head Engine Block Deck Flatness (Measuring the Overall Length of the Cylinder Head):	0.22 mm/0.008 in.
Cylinder Head Exhaust Manifold Deck Flatness:	0.22 mm/0.008 in.
Cylinder Head Intake Manifold Deck Flatness:	0.22 mm/0.008 in.
Cylinder Head Height (Measured from the Cylinder Head Deck to the Valve Rocker Arm Cover Seal Surface):	120.2 mm/4.732 in. (Minimum)

Engine Block

Camshaft Bearing Bore Diameters:	55.063–55.088 mm/2.168–2.169 in.
Engine Block Cylinder Head Deck Surface Flatness	

(Measured within a 152.4 mm/6.0 in.area):	0.11 mm/0.004 in.
Engine Block Cylinder Head Deck Surface Flatness	
(Measuring the Overall Length of the Block Deck):	0.22 mm/0.008 in.
Engine Block Cylinder Head Deck Height	
(Measuring from Centerline of Crankshaft to Deck Face):	234.57–234.82 mm/9.235–9.245 in.
Main Bearing Bore Diameter (Production):	69.871–69.889 mm/2.750–2.751 in.
Valve Lifter Bore Diameter (Production):	21.417–21.443 mm/0.843–0.844 in.
Intake Manifold	
Intake Manifold Cylinder Head Deck Flatness	
(Measured at Gasket Sealing Surfaces):	0.5 mm/0.02 in.
Oil Pan and Front/Rear Cover Alignment	
Oil Pan to Rear of Engine Block Alignment	
(at Transmission Bellhousing Mounting Surface):	0.0–0.25 mm/0.0–0.01 in. (Maximum)
Front Cover Alignment (at Oil Pan Surface):	0.0–0.5 mm/0.0–0.02 in.
Rear Cover Alignment (at Oil Pan Surface):	0.0–0.5 mm/0.0–0.02 in.
Piston	
Piston Outside Diameter (at Size Point):	98.964–98.982 mm/3.8962–3.8969 in
Piston to Bore Clearance (Production):	0.018–0.054 mm/0.0007–0.00212 in
Piston to Bore Clearance (Service Limit):	0.018–0.054 mm/0.0007–0.00212 in. (Maximum)
Piston Pin	
Piston Pin Clearance to Piston Bore (Production):	0.01–0.02 mm/0.0004–0.00078 in
Piston Pin Clearance to Piston Bore (Service Limit):	0.01–0.02 mm/0.0004–0.00078 in.(Maximum)
Piston Pin Diameter:	23.997–24.0 mm/0.9447–0.9448 in
Piston Pin Fit in Connecting Rod:	0.02–0.043 mm/0.00078–0.00169 in. (Interference)
Piston Rings	
Piston Compression Ring End Gap, Production Top:	0.23–0.38 mm/0.009–0.0149 in
Piston Compression Ring End Gap, Production–2:	0.44–0.64 mm/0.0173–0.0251 in.
Piston Oil Ring End Gap, Production:	0.18–0.69 mm/0.007–0.0271 in.
Piston Compression Ring End Gap, Service–Top:	0.23–0.38 mm/0.009–0.01496 in. (Maximum)
Piston Compression Ring End Gap, Service–2nd:	0.44–0.64 mm/0.0173–0.0251 in. (Maximum)
Piston Oil Ring End Gap,Service Limit:	0.18–0.69 mm/0.007–0.0271 in. (Maximum)
Piston Compression Ring Groove Clearance	
(Production–Top):	0.04–0.085 mm/0.00157–0.003346 in.
Piston Compression Ring Groove Clearance	
(Production–2nd):	0.04–0.08 mm/0.00157–0.003149 in.
Piston Oil Ring Groove Clearance (Production):	0.01–0.22 mm.0.0004–0.00866 in.
Piston Compression Ring Groove Clearance	
(Service-Top):	0.04–0.085 mm/0.00157–0.003346 in. (Maximum)
Piston Compression Ring Groove Clearance	
(Service-2nd):	0.04–0.08 mm/0.00157–0.003149 in. (Maximum)

Engine Specifications, continued

Piston Oil Ring Groove Clearance (Service Limit):	0.01–0.22 mm/0.0004–0.00866 in. (Maximum)

Valve System

Valve Lifter:	Hydraulic Roller
Valve Rocker Arm Ratio:	1.70:1
Valve Lash Net Lash:	No Adjustment
Valve Face Angle:	45 degrees
Valve Seat Angle:	46 degrees
Valve Seat Runout:	0.05 mm/0.002 in/ (Maximum)
Valve Seat Width (Intake):	1.02 mm/0.04 in.
Valve Seat Width (Exhaust):	1.78 mm/0.07 in.
Valve Stem Clearance (Production–Intake):	0.025–0.066 mm/0.001–0.0026 in.
Valve Stem Clearance (Production—Intake):	0.025–0.066 mm.0.001–0.0026 in.
Valve Stem Clearance (Production—Exhaust):	0.025–0.066 mm.0.001–0.0026 in.
Valve Stem Clearance (Service—Intake):	0.093 mm/0.0037 in. (Maximum)
Valve Stem Clearance (Service—Exhaust):	0.093 mm/0.0037 in/ (Maximum)
Valve Stem Diameter (Production):	7.955–7.976 mm/0.313–0.314 in.
Valve Stem Diameter (Service):	7.9 mm/0.311 in. (Minimum)
Valve Spring Free Length:	52.9 mm/2.08 in.
Valve Spring Pressure (Closed):	340 N at 45.75 mm/76 lb at 1.80 in.
Valve Spring Pressure (Open):	980 N at 33.55 mm/220 lb at 1.32 in.
Valve Spring Installed Height (Intake):	45.75 mm/1.8 in.
Valve Spring Installed Height (Exhaust):	45.75 mm/1.8 in.
Valve Lift (Intake):	11.99 mm/0.472 in.
Valve Lift (Exhaust):	12.15 mm/0.479 in.
Valve Guide Installed Height (Measured from the Cylinder Head Spring Seat Surface to the Top of the Valve Guide):	17.32 mm/0.682 in.
Valve Stem Oil Seal Installed Height (Measured from the ValveSpring Shim to Top Edge of Seal Body)	18.1–19.1 mm/0.712–0.752 in.

Approximate Fluid Capacities

Cooling System

Automatic Transmission:	11.6 Liters/12.3 quarts
Manual Transmission:	11.9 Liters/12.6 quarts

Engine Crankcase

With Filter:	6.1 Liters/6.5 quarts
Without Filter:	5.7 Liters/6.0 quarts
Fuel Tanks (Total):	72.3 Liters/19.1 gallons

Rear Axle Differential

Lubricant:	1.6 Liters/1.69 quarts	
Limited–Slip Additive	118 Milliliters	4.0 ounces

Transmission Fluid

Drain & Fill (Automatic Transmission):	4.7 Liters/5.0 quarts
Overhaul (Automatic Transmission):	10.2 Liters/10.8 quarts
Overhaul (Manual Transmission):	3.9 Liters/4.1 quarts

TORQUE SPECS

Application	lb.-ft.
AC Compressor Bolts:	37
AC Compressor bracket bolts:	37
AC Idler Pulley Bolt:	37
AC Tensioner Bolt:	18
AIR Pipe to Exhaust Manifold Bolt	15
Camshaft Retainer Bolts:	18
Camshaft Sensor bolt:	18
Camshaft Sprocket Bolts:	26
Catalytic Converter Nut:	18
Compression Test:	
Connecting Rod Bolts First Pass:	15
Connecting Rod Bolts Final Pass:	60 degrees
Coolant Temperature Gauge Sensor:	15
Crankshaft Balancer Bolt: (First Pass use old bolt):	240
Crankshaft Balancer Bolt (Second Pass use NEW bolt:	37
Crankshaft Balancer Bolt (Final Pass):	140 degrees
Crankshaft Bearing Cap Bolts (Inner bolts (1-10) First Pass in sequence):	15

Sequence:
29–25–21–23–27 Side
19–15–11–13–17 Outer
09–05–01–03–07 Inner
Front
10–06–02–04–08 Inner
20–16–12–14–18 Outer
30–26–22–24–28 Side
20 15

Crankshaft Bearing Cap Bolts, inner bolts (1-10) Final Pass in sequence See Crankshaft Bearing Cap Bolts for Sequence:	80 degrees
Crankshaft Bearing Cap Studs (Outer studs (11-20) First Pass in Sequence) See Crankshaft Bearing Cap Bolts for Sequence:	15
Crankshaft Bearing Cap Studs (Outer studs (11-20) Final Pass in Sequence):	53 degrees
Crankshaft Bearing Cap Side Bolts(21-30) See Cranks Bearing Cap Bolts for Sequence:	18
Crankshaft Oil Deflector Nuts:	18
Crankshaft Position Sensor Bolt:	18
Cylinder Head Bolts (First Pass all M11 bolts in sequence):	22
Cylinder Head Bolts (Second Pass all M11 bolts in sequence):	90 degrees
Cylinder Head Bolts (Final Pass all M11 bolts in sequence, excluding the Medium Length bolts at the front and rear of each cylinder head):	90 degrees
Cylinder Head Bolts (Final Pass medium length M11 bolts at the front and rear of each cylinder head):	50 degrees
Cylinder Head Bolts (M8 Inner Bolts in sequence):	22
Cylinder Head Coolant Plug:	15
Cylinder Head Core Hole Plug:	15
Drive Belt Idler Pulley Bolt:	37
Drive Belt Tensioner Bolts:	37
Engine Block Coolant Drain Plugs:	44
Engine Block Heater:	30
Engine Block Oil Gallery Plugs:	44
Engine Crossmember Bolts Large:	107
Engine Crossmember Bolts Small:	92
Engine Flywheel Bolts First Pass (Sequence: 1–4–6–2–3–5):	15
Clutch Pressure Plate Bolts Three passes (Sequence: 1–4–6–2–3–5):	20, 40, 52
Engine Flywheel Bolts Second Pass:	37
Engine Flywheel Bolts Final Pass:	74
Engine Flywheel to Torque Converter Bolt:	44
Engine Front Cover Bolts:	18
Engine Mount Heat Shield Nuts:	89 inch-lbs.
Engine Mount Stud-to-Engine Block:	37
Engine Mount Through Bolts:	70
Engine Mount Through Bolt Nuts:	59
Engine Mount to Engine Block Bolts:	37
Engine Rear Cover Bolts:	18
Engine Service Lift Bracket M10 Bolts:	37
Engine Service Lift Bracket M8 Bolt:	18
Engine Valley Cover Bolts:	18
EGR Valve Bolts First Pass:	89 inch-lbs.
EGR Valve Bolts Final Pass:	22
EGR valve Pipe to cylinder head Bolts:	37
EGR valve Pipe to Exhaust manifold Bolts:	22
EGR valve Pipe to Intake manifold bolt:	89 inch-lbs.
Exhaust Manifold Bolts (First Pass)	

Sequence: 6-3-1-2-4-5:	11
Exhaust Manifold Bolts (Final Pass):	18
Exhaust Pipe Manifold nuts:	26
Front Shock to Engine Crossmember bolt:	48
Fuel Injection Fuel rail bolts:	89 inch-lbs.
Generator Bracket Bolts:	37
Generator rear bracket to engine block bolt:	18
Generator rear bracket to Generator Bolt:	18
Ground Strap Bolt (Rear of cylinder head):	37
Ignition Coil Bolts:	106 inch-lbs.
Ignition Coil Wire Harness Connector Bolts:	106 inch-lbs.
Intake Manifold Bolts (First Pass in sequence):	44
Intake Manifold Bolts (Final Pass in sequence):	89 inch-lbs.
Knock Sensors:	15
Oil Filter:	22
Oil Filter Fitting:	40
Oil Level Indicator Tube Bolt:	18
Oil Level Sensor:	115 inch-lbs.
Oil Pan Baffle Bolts:	106 inch-lbs.
Oil Pan Close out Bolt left side:	106 inch-lbs.
Oil Pan Close out Bolt right side:	106 inch-lbs.
Oil Pan Drain Plug:	18 inch-lbs.
Oil Pan M8 Bolts Oil (pan to engine block and oil pan to front cover):	18
Oil Pan M6 Bolts (Oil pan to rear cover):	106 inch-lbs.
Oil Pressure Sensor:	15
Oil Pump to engine block bolts:	18
Oil Pump cover bolts:	106 inch-lbs.
Oil Pump Relief Valve Plug:	106 inch-lbs.
Oil Pump Screen nuts:	18
Oil Pump screen-to-oil pump bolt:	106 inch-lbs.
Oil Transfer Cover bolts:	106 inch-lbs.
02 Sensor:	31
PCV System Strap (at right front vapor vent pipe stud):	106 inch-lbs.
Power Steering Pump Bolts:	18
Power Steering Pump Brace Bolts:	18
Power Steering Pump Bracket Bolts:	37
Spark Plugs (Cylinder Head New):	15
Spark Plugs (All subsequent installations):	11
Starter Motor Bolts:	37
Throttle Body Bolts:	106 inch-lbs.
Transmission Bell housing Bolt:	35
Valve Lifter Guide Bolts:	106
Valve Rocker Arm Bolts:	22
Valve Rocker Arm Cover Bolts:	106 inch-lbs.
Vapor Vent Pipe Bolts:	106 inch-lbs.
Water Inlet Housing Bolts:	11
Water Pump Bolts (First Pass):	11
Water Pump Bolts (Final Pass):	22
Water Pump Cover Bolts:	11
Water Pump Pulley Bolts (First Pass):	89 inch-lbs.
Water Pump Pulley Bolts (Final Pass):	18

Information supplied by Eboggs_jkvl per 1998 Helms Manual Volume 3, 6-229-6-231

Chris Endres was born with gasoline in his veins, and has been a diehard car enthusiast for most of his life. He cut his teeth on a string of carbureted musclecars before realizing that the future was with EFI. After a brief fling with Buick Grand Nationals, he stumbled across his first LS1, and they've been together ever since!

A resident of suburban Minneapolis, Chris is a regular tech contributor to such magazines as *Popular Hot Rodding* and *GM High-Tech Performance*. This is his first book.

OTHER BOOKS BY HPBOOKS

HANDBOOKS

Auto Electrical Handbook: 0-89586-238-7/HP1238
Auto Upholstery & Interiors: 1-55788-265-7/HP1265
Car Builder's Handbook: 1-55788-278-9/HP1278
4-Wheel & Off-Road's Chassis & Suspension Handbook: 1-55788-406-4
The Lowrider's Handbook: 1-55788-383-1/HP1383
Powerglide Transmission Handbook:1-55788-355-6/HP1355
Turbo Hydramatic 350 Handbook: 0-89586-051-1/HP1051
Welder's Handbook: 1-55788-264-9/HP1264

BODYWORK & PAINTING

Automotive Detailing: 1-55788-288-6/HP1288
Automotive Paint Handbook: 1-55788-291-6/HP1291
Fiberglass & Composite Materials: 1-55788-239-8/HP1239
Metal Fabricator's Handbook: 0-89586-870-9/HP1870
Paint & Body Handbook: 1-55788-082-4/HP1082
Pro Paint & Body: 1-55788-394-7
Sheet Metal Handbook: 0-89586-757-5/HP1757

INDUCTION

Bosch Fuel Injection Systems: 1-55788-365-3/HP1365
Holley 4150: 0-89586-047-3/HP1047
Holley Carbs, Manifolds & F.I.: 1-55788-052-2/HP1052
Rochester Carburetors: 0-89586-301-4/HP1301
Turbochargers: 0-89586-135-6/HP1135
Weber Carburetors: 0-89586-377-4/HP1377

PERFORMANCE

Baja Bugs & Buggies: 0-89586-186-0/HP1186
Big-Block Chevy Performance: 1-55788-216-9/HP1216
Big-Block Mopar Performance: 1-55788-302-5/HP1302
Bracket Racing: 1-55788-266-5/HP1266
Brake Systems: 1-55788-281-9/HP1281
Camaro Performance: 1-55788-057-3/HP1057
Chassis Engineering: 1-55788-055-7/HP1055
Chevy Trucks: 1-55788-340-8/HP1340
Ford Windsor Small-Block Performance: 1-55788-323-8/HP1323
4Wheel&Off-Road's Chassis & Suspension: 1-55788-406-4/HP1406
Honda/Acura Engine Performance: 1-55788-384-X/HP1384
High Performance Hardware: 1-55788-304-1/HP1304
How to Hot Rod Big-Block Chevys: 0-912656-04-2/HP104
How to Hot Rod Small-Block Chevys: 0-912656-06-9/HP106
How to Hot Rod Small-Block Mopar Engine Revised: 1-55788-405-6
How to Hot Rod VW Engines: 0-912656-03-4/HP103
How to Make Your Car Handle: 0-912656-46-8/HP146
John Lingenfelter: Modify Small-Block Chevy: 1-55788-238-X/HP1238
LS1/LS6 Small-Block Chevy Performance: 1-55788-407-2/HP1407
Mustang 5.0 Projects: 1-55788-275-4/HP1275
Mustang Performance (Engines): 1-55788-193-6/HP1193
Mustang Performance 2 (Chassis): 1-55788-202-9/HP1202
Mustang Perf. Chassis, Suspension, Driveline Tuning: 1-55788-387-4
Mustang Performance Engine Tuning: 1-55788-387-4/HP1387
1001 High Performance Tech Tips: 1-55788-199-5/HP1199
Performance Ignition Systems: 1-55788-306-8/HP1306
Small-Block Chevy Performance: 1-55788-253-3/HP1253
Small Block Chevy Engine Buildups: 1-55788-400-5/HP1400
Stock Car Setup Secrets: 1-55788-401-3/HP1401
Tuning Accel/DFI 6.0 Programmable F.I: 1-55788-413-7/HP1413

ENGINE REBUILDING

Engine Builder's Handbook: 1-55788-245-2/HP1245
How to Rebuild Small-Block Chevy LT-1/LT-4: 1-55788-393-9/HP1393
Rebuild Air-Cooled VW Engines: 0-89586-225-5/HP1225
Rebuild Big-Block Chevy Engines: 0-89586-175-5/HP1175
Rebuild Big-Block Ford Engines: 0-89586-070-8/HP1070
Rebuild Big-Block Mopar Engines: 1-55788-190-1/HP1190
Rebuild Ford V-8 Engines: 0-89586-036-8/HP1036
Rebuild GenV/Gen VI Big-Block Chevy: 1-55788-357-2/HP1357
Rebuild Small-Block Chevy Engines: 1-55788-029-8/HP1029
Rebuild Small-Block Ford Engines: 0-912656-89-1/HP189
Rebuild Small-Block Mopar Engines: 0-89586-128-3/HP1128

RESTORATION, MAINTENANCE, REPAIR

Camaro Owner's Handbook ('67–'81): 1-55788-301-7/HP1301
Camaro Restoration Handbook ('67–'81): 0-89586-375-8/HP1375
Classic Car Restorer's Handbook: 1-55788-194-4/HP1194
How to Maintain & Repair Your Jeep: 1-55788-371-8/HP1371
Mustang Restoration Handbook ('64 1/2–'70): 0-89586-402-9/HP1402
Tri-Five Chevy Owner's Handbook ('55–'57): 1-55788-285-1/HP1285

GENERAL REFERENCE

Auto Math Handbook: 1-55788-020-4 /HP1020
Corvette Tech Q&A: 1-55788-376-9/HP1376
Ford Total Performance, 1962–1970: 1-55788-327-0/HP1327
Guide to GM Muscle Cars: 1-55788-003-4/HP1003

MARINE

Big-Block Chevy Marine Performance: 1-55788-297-5/HP1297
Small-Block Chevy Marine Performance: 1-55788-317-3/HP1317

ORDER YOUR COPY TODAY!
All books can be purchased at your favorite retail or online bookstore (use ISBN number), or auto parts store (Use HP part number). You can also order direct from HPBooks by calling toll-free at 800/788-6262, ext. 1.